Connecting Sub-Saharan Africa

A World Bank Group Strategy for Information and Communication Technology Sector Development

Pierre Guislain
Mavis A. Ampah
Laurent Besançon
Cécile Niang
Alexandre Sérot

THE WORLD BANK
Washington, D.C.

ISBN-10: 0-8213-6150-3   ISBN-13: 978-0-8213-6150-4
eISBN: 0-8213-6161-1
ISSN: 1726-5878   DOI: 10.1596 /0-8213-6150-4

Pierre Guislain is Manager of the Policy Division in the Global Information and Communications Technologies Department of the World Bank. Mavis A. Ampah is Senior ICT Policy Specialist in the same division. Laurent Besançon is Regulatory Specialist in the same division. Cécile Niang is Operations Analyst in the same division. Alexandre Sérot is from the Global Information and Communication Technologies Department, a joint department of the World Bank and the IFC.

Library of Congress Cataloging-in-Publication Data
Connecting Sub-Saharan Africa: a World Bank Group strategy for information and
    communication technology sector development / Pierre Guislain ... [et al.].
        p. cm. — (World Bank working paper; no. 51)
        Includes bibliographical references.
        ISBN 0-8213-6150-3 (pbk.)
        1. Information technology—Africa, Sub-Saharan. 2. Telecommunication—Africa,
Sub-Saharan. 3. Economic assistance—Africa, Sub-Saharan. 4. World Bank—Africa,
Sub-Saharan. I. Guislain, Pierre, 1957- II. World Bank Group. III. World Bank.
IV. Series.

HC 800.Z9I5538 2005
384'.096—dc22

2005045768

Contents

Contents

Contents

Appendixes

LIST OF BOXES

LIST OF FIGURES

Foreword

Information and Communication Technology (ICT) has an essential role to play in poverty alleviation and provides powerful tools to achieve the Millennium Development Goals. No longer are information and communication networks and services luxuries for developing countries—they are now a necessity. ICT gives private sector enterprises access to new opportunities for trade, helps to diversify economies and spurs economic development. It enables educators to network and learn, and develop new and better techniques for reaching students. ICT also enables governments and civil society organizations to coordinate their fight against poverty, and facilitates activities ranging from delivery of critical health services and promoting civic participation to fostering innovation and entrepreneurship.

While Africa has traditionally lagged behind other regions in the development of ICT infrastructure and services, today it has moved to the forefront of the continent's development agenda. ICT's prominence is reflected by the focus given by Africa's heads of state to the World Summit on the Information Society, and by the ICT Priority Projects of the New Partnership for Africa's Development (NEPAD). Countries are clearly recognizing that ICT can help to address some of the region's challenges by providing the underlying means for new interactions within any given society or across the region that can in turn facilitate social and economic development.

In this context, the World Bank Group has established a revitalized ICT agenda for the region. Its *Africa Region Development Strategy,* adopted in July 2003, identifies ICT as one of the three emerging positive factors of the 21st century for Africa, offering enormous opportunities to leapfrog stages of development.

ICT can and should play a significant role in Africa's efforts to drive strong regional development and integration in the 21st century. The World Bank Group has supported many of the region's reform programs that have been instrumental in driving the phenomenal growth in connectivity that the region has witnessed. These successes, however, should not overshadow the critical gaps that still exist in terms of broadening access to the wider population of Africa, ensuring affordable prices, and mainstreaming the use of ICT for development.

What then are the new challenges that need to be addressed? *Connecting Sub-Saharan Africa* outlines a strategy for ICT development in Sub-Saharan Africa that will further the reform agenda to facilitate information and communication infrastructure deployment and creation of ICT applications. This report was prepared by a team from the Global Information and Communication Technologies Department of the World Bank, drawing on the group's experience in covering active policy dialogue across the African continent.

We hope that this paper will help demonstrate the importance of having an ICT strategy and roadmap as an integral part of the Sub-Saharan Africa development agenda, and the need for the development community to support African governments in this regard.

Mohsen A. Khalil
Director
Information and Communication Technologies Department

Acknowledgments

This Africa Region ICT Strategy was written by a team led by Pierre Guislain, and including Mavis Ampah, Laurent Besançon, Cécile Thioro Niang and Alexandre Sérot (GICT). A special thanks is given to Jared Green (Consultant) for his help in writing the report. Valuable contributions and guidance were received from Mohsen Khalil, (GICT), Marie Françoise Marie-Nelly (AFRCE), Demba Ba and Hervé Assah (AFTPS). Valuable assistance to prepare the report for publication was provided by Andrea Ruiz-Esparza and Mark Andrew Wahl (GICT). Additional inputs were provided by Boutheina Guermazi, Jérôme Bezzina, Isabelle Huynh-Segni, Yann Burtin, Irene Christy, Charles Kenny, Christine Zhen-Wei Qiang, Juan Navas-Sabater, Tenzin Norbhu, Michele Rajaobelina, Robert Stephens, Svetoslav Tintchev, Kofi Arkaah, Solomon Asamoah, Stephanie Von Friedeburg, Joseph Solan, Darren Massara, Stephen Chow, Vera Lapshina and Ravi Vish (GICT), Mostafa Terrab, Seth Ayers and Kerry McNamara (*info*Dev), Abir Burgul and Nabil Fawaz (MIGA), Paul Noumba (WBI), and Bobak Rezaian (AFTQK).

Acronyms and Abbreviations

$	U.S. dollar (unless otherwise noted)
AVU	African Virtual University
COMESA	Common Market for Eastern and Southern Africa
EASSy	East Africa Submarine Cable System
ECOWAS	Economic Community of West African States
E&SA	Eastern and Southern Africa
GDLN	Global Development Learning Network
ICT	Information and communication technologies
ISP	Internet service providers
IT	Information technology
ITU	International Telecommunication Union
IXP	Internet exchange point
NEPAD	New Partnership for Africa's Development
OBA	Output-based aid
SADC	Southern African Development Community
SSA	Sub-Saharan Africa
VSAT	Very small aperture terminal
WBG	World Bank Group

Executive Summary

With a revitalized approach to information and communication technologies (ICT) sector development in Sub-Saharan Africa (SSA), the World Bank Group (WBG) is poised to support further reform of the telecommunications sectors in the region. The strategy builds on the earlier reform agenda in the sector by leveraging the strengths of the Bank's various organizational units to advance the essential goal of increasing the continent's connectivity. Of particular concern is the ability to bring rural areas into the national, regional, and global economies, thus creating new opportunities for the world's poorest citizens.

In its renewed efforts, the World Bank is especially keen on developing and enhancing the capacity of Africa's ICT institutions—including regulators, ministries, and regional bodies—to lead the development of an interconnected region and implement sustainable regional strategies for integration and knowledge sharing. The WBG is exploring critically important new ways to strengthen support for local ICT applications which will help to create the requisite skills needed to adapt technologies to SSA's circumstances and enable Africans to create innovative solutions to their own development challenges.

Connecting Sub-Saharan Africa outlines a strategy for WBG involvement in SSA. It is composed of three pillars, all of which underpin a robust regional integration and connectivity agenda, as depicted in the figure below:

The Three Core Pillars of the WBG ICT Strategy for SSA

The Core Reform Agenda	**Addressing Market Failures**	**ICT for Development Applications**
- Market liberalization - Regulation - Capacity-building - Privatization - Postal sector reform	- Rural access - National backbone - Post-conflict countries	- E-commerce - E-government - Civil society applications

Regional Integration and Connectivity

Pushing Forward the Core Reform Agenda Remains a Priority

First, the WBG will continue to consolidate telecommunications sector reforms made over the past decade. Reform is required to further expand access to infrastructure and services at more competitive prices.

A great majority of SSA countries have implemented various aspects of the core reform agenda, but have not yet completed the process. The WBG and other development partners need to retain a focus on pursuing the implementation of the "core reform agenda," as well as assisting in formulation (or revision) and implementation of pragmatic liberalization, privatization, and postal sector development strategies that fit national circumstances. This involves creating and strengthening autonomous regulators in SSA, given their crucial role in implementing competition policies and monitoring market progress, as well as building the capacity of ministries to participate in national reform and regional harmonization efforts. More generally, the development community needs to increase efforts to assist governments to develop pro-competitive legal and regulatory frameworks required to create a competitive marketplace within SSA, in both the telecommunications and postal sectors. Reformed postal sectors will support SSA's overall social and economic development agenda through their rural network and must continue to be modernized in the wake of reforms in the telecommunications sector.

Pushing forward this core reform agenda is the foundation for sustainable ICT sector development and rollout of its infrastructure. Only when effective access to infrastructure is made available will it be possible to engage in substantive discussion about developing "ICT for development" applications such as e-commerce, e-government, and civil society-focused applications. Significant evidence exists to support the fact that liberalized markets based on pro-competitive policies and regulatory frameworks provide strong support for improved access.

Addressing Market Failures

Second, the WBG will intensify its effort to address market failures by helping countries devise innovative private—public sector partnerships to close infrastructure and service gaps in key areas.

Even with the successful implementation of the core reform agenda, additional government support may be required to create incentives for private sector investment in key areas, including: providing access to rural and underserved areas; developing the embryonic national backbone; and deploying broadband, government networks, and cross-border connectivity projects. The WBG and its development partners should support SSA governments in instances of proven market failure, and in situations where competitive forces have been unable to achieve the desired policy objectives. It is critical, however, that public support does not distort competition in growing ICT sectors—policy and regulatory interventions that can influence market development should be explored before public financing solutions are applied. For instance, the introduction of output-based aid (OBA) schemes for rural access or backbone infrastructure development in less reformed environments is a risky proposition, unless it is preceded by detailed impact analysis to avoid subsidizing what could otherwise be commercially viable operations if the regulatory environment was set up correctly.

Addressing the special telecommunications needs of post-conflict countries is also critical to bring these countries out of crisis, and to resume the promotion of investments. To facilitate the involvement of the development community in post-conflict SSA countries, additional research needs to be conducted to assess ongoing post-conflict intervention projects in the ICT sector. While financial instruments must be identified to remove constraints to ICT development in post-conflict countries, strong focus should be directed on strengthening the regulatory environment to secure the new flow of investments.

Promoting "ICT for Development" Applications

Third, the WBG will intensify its support to SSA countries in developing the policy and regulatory frameworks that support "ICT for development" applications, which enable economic and social development, and foster the development of infrastructure in key areas.

African policy-makers need to create the policy and regulatory conditions for e-commerce and online financial transactions, while also considering the rapid convergence of technologies. Governments also need to develop e-government initiatives and support the expansion of access so that civil society organizations can leverage the capacities that ICT brings once basic infrastructure issues have been addressed. Incorporating ICT more fully into governments, private sector enterprises, and civil society organizations will help to increase the diffusion and use of applications throughout SSA.

The WBG is developing strategies in a small group of countries to monitor and evaluate national ICT strategies. The main emphasis is to develop common sense approaches to designing ICT strategies that can be implemented, and to assess their success. However, while national ICT strategies may be useful tools for articulating national programs and for mobilizing funds for "ICT for development" applications, the development community can complement efforts by assisting in the review of regulatory and policy impediments to implementing such applications.

A growing number of SSA countries are seeking, and in many cases securing, WBG assistance to develop and implement sector-based ICT programs, including e-education, e-health, and e-government to foster economic development. However, without an overall coordination mechanism in place at the country-level, each of these ICT programs may lead to the build-out of sector-specific ICT infrastructure (for example, national or metropolitan networks) owned and operated by a ministry or other government agency, duplicating existing private sector ICT infrastructure, or infrastructure built by other ministries. It will be critical for SSA countries, with support from development partners, to approach investments in "ICT for development" applications as part of a holistic national strategy for ICT sector development. Governments also need to ensure that investment in ICT applications spurs the growth of national ICT sector industries with the capacity to maintain and grow local applications to serve development objectives.

Accelerating Regional Integration and Connectivity

The World Bank will support the acceleration of regional ICT integration across SSA as a means to develop larger, more viable regional ICT markets, and to foster economic integration. Additionally, the WBG and its development partners should increase efforts to support the regional connectivity initiatives to address cross-border connectivity gaps in the region.

Policy and regulatory harmonization is required for the success of any initiative linking the SSA region. The WBG is already supporting efforts to identify and remove bottlenecks to roaming, to cross-border connectivity, and to international gateway access. Ongoing initiatives help to identify priority cross-border links for which effective public-private partnerships and international donor financing could play a valuable role. Additional work is being done with the support of the World Bank Africa Regional Integration Department to help the region rationalize the various infrastructure projects—underway or in the planning stages—in order to ensure a coherent rollout of regional infrastructure. As part of this effort, the WBG is working with the New Partnership for Africa's Development (NEPAD) to develop a detailed assessment of the various connectivity projects.

Beyond policy and regulatory considerations, it is increasingly apparent that private capital alone may not be sufficient to fill the cross-border connectivity gap. The WBG will explore the need to collaborate with partners in the donor community on the investment in cross-border infrastructure and national backbones for coastal and landlocked countries. These countries would need to ensure open and competitive access to regional infrastructure and/or transparent and nondiscriminatory infrastructure-sharing arrangements.

CHAPTER 1

Introducing the World Bank Group's ICT Strategy for Sub-Saharan Africa

The increasing significance of the role of ICT—particularly the underlying telecommunications infrastructure—in fostering economic and social development is well recognized. ICT is essential to growth and overall country competitiveness; it is a major enabler with the potential to facilitate trade diversification and intensification, as well as to integrate firms in global markets and supply chains. ICT contributes to poverty reduction by increasing productivity and creating new business opportunities, and provides opportunities to make African governments, private sector enterprises, and civil society organizations more effective. More importantly, ICT also serves as a critical driver for reaching key Millennium Development Goals—a current focus for all developing countries.

ICT is a broad and encompassing term that refers to the range of technologies that facilitate the sharing of knowledge. ICT includes: (a) telecommunications infrastructure that provides the underlying foundation for communication through mobile and fixed-line telephony; (b) global distributed networks such as the Internet; and (c) ICT-based industries that include a range of Internet-related and information technology (IT) based industries. The global and national industries that rely on ICT infrastructure and the Internet make up the knowledge economy.

Over the last decade, the impact of ICT use in Africa has been notable. Thanks to sector reforms implemented in SSA, more Africans are using telephones, e-mail, and the Internet to communicate, to share information, to collaborate, to entertain, and to work. Private sector enterprises are using ICT to participate in online market transactions, to capture trade opportunities, and to reap productivity gains. Governments are bringing efficiency to the delivery of public services, and capitalizing on IT-based management systems to improve governance and accountability. Civil society organizations, including those focused on HIV/AIDS prevention, are using ICT to leverage limited resources, delivering educational materials, telemedicine services, and creating greater awareness of social and health issues. Postal

1

systems, as a result of telecommunications sector reform, are modernizing, reorienting, and offering commercial services made possible only through ICT. In a few and promising instances ICT is creating new opportunities for Africa's poorest citizens, by increasing their ability to access agricultural market information and negotiate better prices for their goods.

To help Sub-Saharan Africans connect more efficiently with one and other and the world, the WBG has supported the promotion of competition, the development of regulatory agencies and frameworks, the clear separation of postal and telecommunications activities, the privatization of incumbent telecommunications operators, and the strengthening of telecommunications ministries. This "core reform agenda" has unleashed competitive forces in SSA telecommunications sectors and fostered private sector participation in the fixed and mobile phone, as well as Internet service provider (ISP) markets. As a result, access has improved dramatically and new services have been brought to more people. Thanks to an expansion of the mobile telephone market following some preliminary market reforms, SSA is estimated to have reached 5.8 percent total telephone penetration at the end of 2003 (up from 1 percent in 1990).

However, there is also a need to accelerate the efforts to expand access, particularly in rural areas. The significant and perhaps growing divide between rural and urban ICT access in SSA countries can only be remedied with innovative public-private partnerships that encourage increased investments by the private sector. In addition, Internet access falls well below the world benchmarks—recent statistics show that less than 1 percent of Africans in SSA can access the Internet. The WBG retains its focus on helping SSA countries to address instances of market failure in telecommunications sector development, on creating demand for ICT through innovative applications, and on promoting the development of a regional market. A multifaceted approach aimed at expanding access is required to spread the benefits of ICT in SSA.

"Connecting Sub-Saharan Africa" is a response to the needs of the SSA region, and presents a revitalized approach to ICT development. The strategy outlines a clear vision for WBG involvement in SSA, and is composed of three pillars, all of which underpin a robust regional integration and connectivity agenda as shown in the figure below:

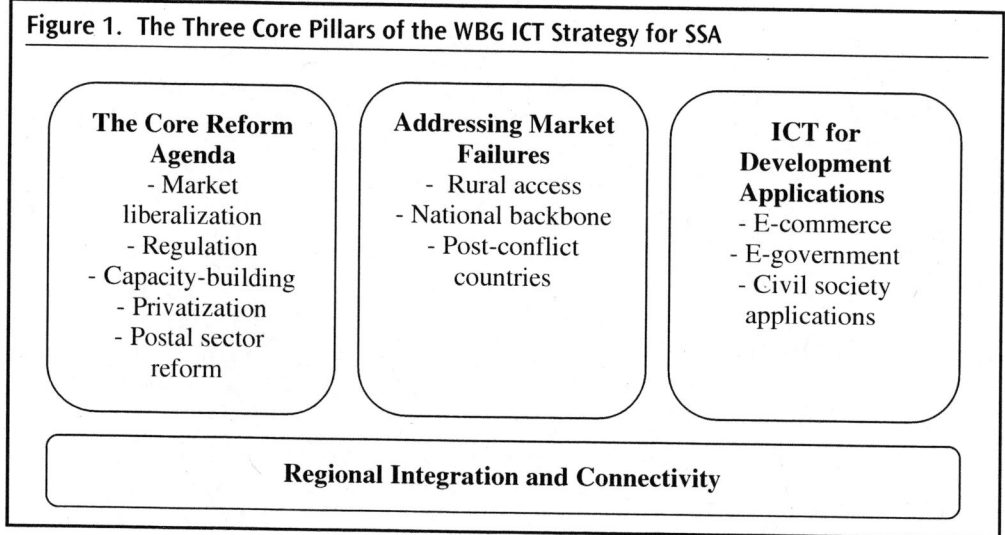

Figure 1. The Three Core Pillars of the WBG ICT Strategy for SSA

The Core Reform Agenda
- Market liberalization
- Regulation
- Capacity-building
- Privatization
- Postal sector reform

Addressing Market Failures
- Rural access
- National backbone
- Post-conflict countries

ICT for Development Applications
- E-commerce
- E-government
- Civil society applications

Regional Integration and Connectivity

The chapters below outline this strategy. Chapter 2 presents the benefits of telecommunications sector reform and provides the rationale for the focus on telecommunications sector reform. Chapter 3 reviews the WBG's strategies for supporting telecommunications sector reform in SSA. Chapter 4 outlines the way the WBG will continue to assist SSA governments to address market failures that cannot be tackled alone by core sector reforms. Chapter 5 expands the scope of the paper into applications, and the ways the WBG will assist SSA governments in developing of "ICT for development" applications. Finally, Chapter 6 discusses the WBG's plans to assist regional organizations to foster regional ICT integration and connectivity.

CHAPTER 2

The Benefits of Telecommunications Sector Reform

The linkages between ICT, growth and competitiveness make it all the more relevant to tackle the core reform agenda in the short-term. Core sector reform, as supported by the World Bank, has already triggered significant private investment in networks and services throughout the SSA—which over the last decade has translated into a near six-fold increase in overall telephone penetration. Even so, Africa lags behind in providing universal access, and runs the risk of remaining marginalized and outside of the global ICT revolution, unless there is a concerted effort to accelerate the pace of reforms and to consolidate achieved results.

Reform Increases Access to Voice Services

The introduction of competition, private sector participation, and the development of regulatory frameworks has led to better ICT access in SSA. The number of telephones per capita increased from about 1 percent in 1990 to 5.8 percent in 2003 (or from 0.37 percent to 3.4 percent, excluding South Africa). In aggregate terms, the mobile market has grown from under 20,000 users in 1993 for SSA (excluding South Africa) to an estimated 18.2 million users in 2003. The phenomenal uptake of mobile services largely explains the rapid growth of access to voice services over the past decade.

While there are significant gains made in many SSA countries, the cost of failing to reform has also increased given current investment constraints. As countries delay implementing the core reform agenda, they also continue to fail to attract viable investments into the sector to support the expansion of access. In a depressed global investment climate for telecommunications, SSA countries increasingly compete for investment funds with other developing regions of the world. The cases of Ethiopia and Mauritania are good illustrations

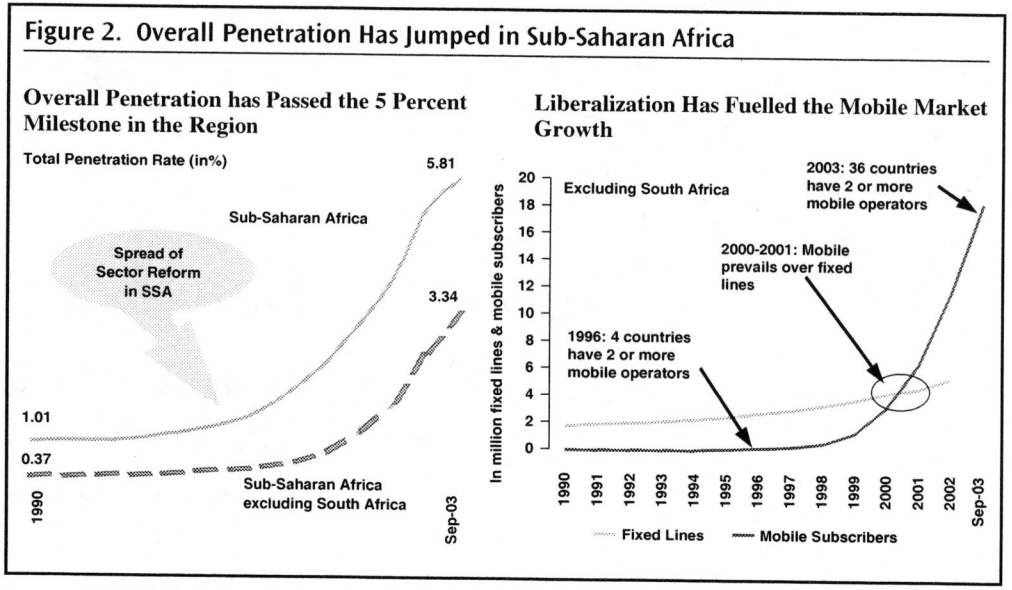

Figure 2. Overall Penetration Has Jumped in Sub-Saharan Africa

Source: WBG Analysis based on International Telecommunications Union (ITU) World Telecommunications Indicators Database, ITU World Telecommunication Regulatory Database, WDI and Middle East and Africa Wireless Analyst for 2003 Data on Mobile.

of the impact of reform on access. In Mauritania recent reforms have helped its move in the ranking from 27th to 10th among SSA countries—in terms of telephone penetration between 1995 and 2003—by contrast, Ethiopia, as a result of a continued lack of reform, has dropped to the bottom in the rankings from 39th to 48th (Appendix B).

Reform Expands Internet Access

Opening the ISP market to competition has led to an increase of the number of Internet users in the region. ISPs—wholesale and retail sellers of Internet access—now provide service via fixed-line and satellite connections across SSA. As a result, Internet access has increased considerably. There are an estimated 3.2 million Internet users in SSA (excluding South Africa) at year-end 2002, up from 0.2 million in 1998.[1] However, user penetration still stands at 93 per 10,000 users for SSA (51 if South Africa is excluded) compared to 982 for the world—one-tenth of the world's average.

For SSA to enjoy the wealth of information accessible via the Internet, and to benefit from the trade and commercial opportunities provided by Internet access, additional infrastructure must be developed, and stronger competitive pressures must be introduced in the ISP market. Only further reforms aimed at the ISP and broadband markets can spur the development of the infrastructure.

---

1. ITU, "Birth of Broadband," 2003 Internet Report; and ITU, *1999 World Telecommunications Development Report.*

Box 1. The Case for Reform: A Comparison of Mauritania and Ethiopia

In 1998, Mauritania embarked on an ambitious reform of its telecommunications sector. The WBG supported the reform process through financing consultancy services, transaction advisors, and efforts in capacity-building. These helped to set up a sound legal and regulatory framework, to establish a fully operational and effective regulatory agency, to open the market to competition, and to privatize the incumbent operator. The reform program brought major benefits to users by improving access to services and lowering prices. Overall teledensity jumped from 0.6 percent in 1998, to 11.07 percent in 2003. A credible and transparent sectoral framework also attracted record levels of private investment and rapid rollout: proceeds from the sale of two mobile licenses totaled $56 million, while the two operators launched services in a record six months. Privatization proceeds from the fixed operator amounted to an additional $48 million (equivalent to $4,065 per line), of which more than $32 million was injected as fresh capital into the company through a capital increase. The macroeconomic impact of the reform goes well beyond these one-off revenues: the continuous development of the sector has led to higher fiscal revenues, the creation of a large number of new jobs, and increased global competitiveness.

Ethiopia on the other hand, has not undertaken substantial reform of its telecommunications sector. Its current policy of maintaining a monopoly in the provision of all telecommunications infrastructure and services has cost the nation in inefficiency and opportunity costs. The government has yet to articulate a sector reform strategy and liberalization timetable. The lack of reform has manifested poor sector growth and performance. In fact, whereas telecommunication penetration in Mauritania increased from 0.41 percent in 1995, to 11.07 percent in 2003 as a result of sector reform (moving in the rankings from the 27th to the 10th in Africa, out of 48 countries), in Ethiopia, where sector reform has been lagging, the increase was only from 0.25 percent to 0.61 percent over the same period (bringing the country from 39th to 48th place).

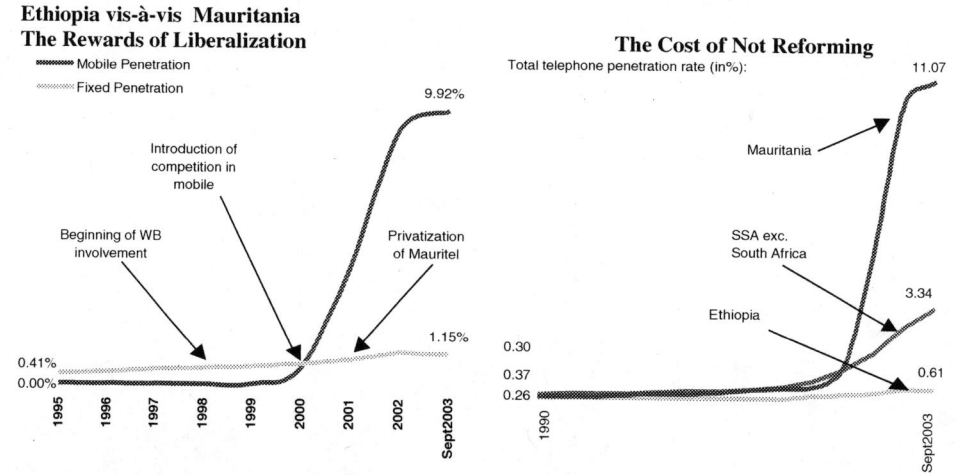

Ethiopia vis-à-vis Mauritania
The Rewards of Liberalization

The Cost of Not Reforming

Source: WBG GICT analysis based on ITU World Telecommunications Indicators Database, WDI and Middle East and Africa Wireless Analyst for 2003 Data on Mobile.

Reform Reduces the Cost of Conducting Business

Competition leads to lower prices, as well as to improved quality and availability of services. Currently, SSA pays the highest price in the world for international calls. Similarly, the price of broadband access is also highest in SSA. Introducing competition in the

Figure 3. Sub-Saharan Africa Lags Behind

	Total Penetration (in%)			Internet 2002		
	2002	**1995**	**CARG 95-02**	**Estimated Internet Users per 10.000 inhab.**	**Internet hosts per 10.000 inhab.**	**Number of PC per 100 inhab.**
WORLD AVERAGE	36.5	13.78	15%	981.79	235.91	9.26
SSA AVERAGE	5.26	1.25	23%	93.04	3.48	1.05
SSA AVERAGE Exc. SA	2.69	0.48	28%	50.66	0.58	0.68

Note: (a) 2001 Data for Mainline in Liberia and Mayotte, (b) 2001 Data for Mobile in Liberia for

WB Analysis based on ITU World Telecommunications Indicators Database

provision of broadband services and international telephone services is critical to support the development of a globally competitive African business sector that can connect with other businesses around the world and communicate at a low cost. Lower prices for international calls are highly correlated with the level of competition. The availability of high-quality services is also a primary concern for business users.

Figure 4. Lack of Competition Leads to Higher Prices

Lack of Competition Leads to High Prices for International Calls in Sub-Saharan Africa

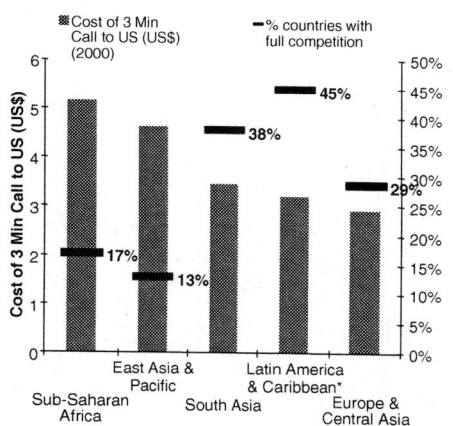

Africa Has the Most Expensive Broadband Costs, Hindering the Development of Internet Access

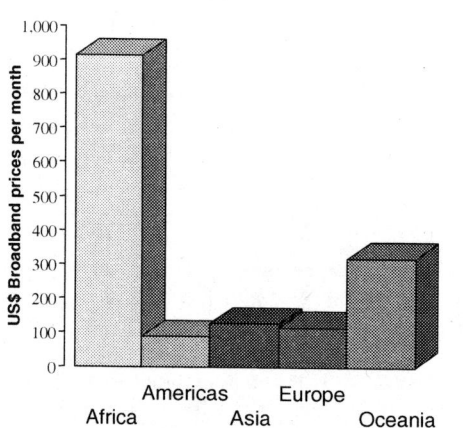

Notes: (a) Prices as of July 2003; (b) ITU calculation for Africa is based on a limited number of countries due to scarcity of data.

Based on data from Birth of Broadband, September 2003, ITU

:* LAC data are for 1999.
The MENA region has no countries with full competition.

Competition in International Voice Communications. World Bank, 2003, based on World Development Indicators, citing International Telecommunications Union data.

CHAPTER 3

Pushing Forward the Core Reform Agenda

The ICT sector is often the first infrastructure sector to go through a phase of profound reforms, with the introduction of private sector participation through competition, and the privatization of incumbent telecommunications operators. Reform also involves the establishment of pro-competitive policy and regulatory frameworks, as well as the creation of regulatory bodies to promote competition, and balance the interests of users with those of the private sector (see Appendix B for the status of sector reform in SSA). Figure 5 shows that the core sector reform agenda has been unevenly implemented in SSA countries.

Thus far in SSA most liberalization efforts have concentrated on the provision of mobile services. These efforts must be generalized to unleash the potential of fixed-line and international market segments, as well as Internet and broadband markets. In many countries incumbent operators still enjoy a monopoly of basic voice telephony over fixed-lines, international services, and the provision of infrastructure for data and other Internet related services. Competition is still not allowed, or is restricted in a majority of countries.

In Africa the privatization of state-owned incumbents has been slow to take hold for both political and market reasons. New approaches need to be explored to deal with state-owned incumbents, such as introducing management contracts and other alternatives. Success in the privatization of state-owned incumbents requires a firm grounding of transactions in a dynamic and pro-competitive environment.

More broadly, telecommunications sector reform also creates new challenges for postal sectors in SSA, which are state-owned. Swift reform in the telecommunications sector often results in postal sectors being left behind. Postal sectors now need to undergo strategic reorientation, to move towards a more corporate approach to delivering postal and other services to citizens, as well as to delivering value-added services to governments and the private sector.

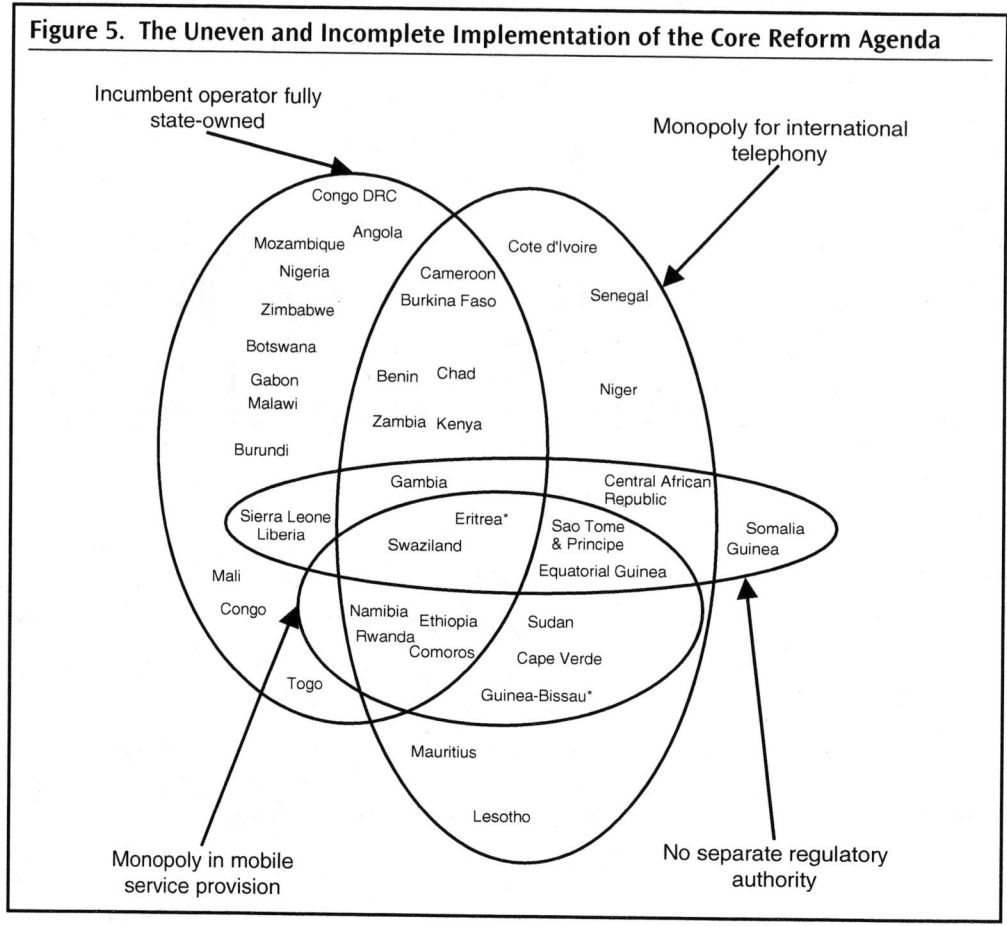

Figure 5. The Uneven and Incomplete Implementation of the Core Reform Agenda

Source: WBG GICT analysis, 2004.

The countries of Ghana, Madagascar, Mauritania, Seychelles, South Africa, Tanzania, and Uganda are considered champions of successful telecommunications sector reforms—reforms that have brought significant changes in terms of access and quality of service. In Mauritania, teledensity jumped from 0.6 percent at the time of the reform in 1998, to 11.07 percent in 2003. However, even among these champions the privatization process is not completed, competition in international voice traffic remains partial, and interconnection disputes still hamper the development of the sector.

Many SSA countries have a long way to go towards reforms, and pursuing the core reform agenda should remain a priority. The WBG will continue to provide support for policy and regulatory capacity-building to create, or maintain, an environment conducive to private investment. WBG assistance will continue to:

■ *Strengthen the regulatory agencies' capacity* to promote competition and to regulate the sector where needed; develop the right mix of expertise to implement pragmatic regulatory strategies involving transparent and participatory regulatory processes; benefit from international best practice; and access rapid intervention instruments when key expertise is needed to address urgent regulatory issues.

▧ *Support sector ministries' capacity* to formulate clear sector policies aimed at promoting cost-effective, widespread, and pro-competitive deployment of infrastructure and services; to become an active partner in regional harmonization and integration initiatives initiated by regional economic communities; and to monitor the overall progress of sector reform.

▧ *Promote competition through pragmatic market liberalization strategies beyond mobile services,* thus combining competitive forces with appropriate levels of regulation to leverage the most powerful instrument to improve access, to reduce prices, and to foster sector development.

▧ *Explore new reforms and pragmatic privatization approaches, as well as regulatory tools* for the incumbent operator (in a difficult but improving global telecommunications environment) to allow corporate reform while promoting a level playing field between all operators (public or private).

▧ *Support pragmatic postal sector reforms through legal and regulatory approaches,* which foster conditions to liberalize, commercialize, or to introduce private-sector participation and partnerships into the postal sector, and encourage the development of new forms of postal businesses.

Most countries in SSA are faced with the same broad ICT sector challenges, namely, starting or completing the basic reform agenda. Currently the WBG supports about 30 SSA countries with ICT policy and regulatory reform.

Strengthening Regulatory Frameworks

Though the SSA region scores well on the set-up of regulatory agencies, a recent Public Private Investment Advisory Facility study for the Economic Community of Western Africa (ECOWAS) found that, in practice, 80 percent of the newly created agencies were not deemed autonomous—autonomous being defined as the power to both *make* and *enforce* regulations.

Another study, commissioned by the World Bank's Global Information and Communication Department, to assess the effectiveness of regulators in the Africa region, found that the lack of financial and human resources seriously hinders the ability of regulatory agencies to be responsive to market developments, to develop independent expertise, to retain performing staff, and to build their credibility in the sector.[2] In addition, the study found that the lack of clear and enforced functional separation of responsibilities between the regulatory agency and the policymaker (as well as other agencies) impedes the effectiveness of the regulators, and ultimately the development of the sector.

The ability of countries to attract private investment is deeply conditioned by the overall quality of the regulatory framework. Access to private investment is required on the occasion of the transfer of ownership of incumbent, sales of stake, market liberalization opportunities, or for the expansion of existing activities. Access to financing can be improved

---

2. Framework for evaluating the effectiveness of telecommunications regulators in SSA, NERA (forthcoming).

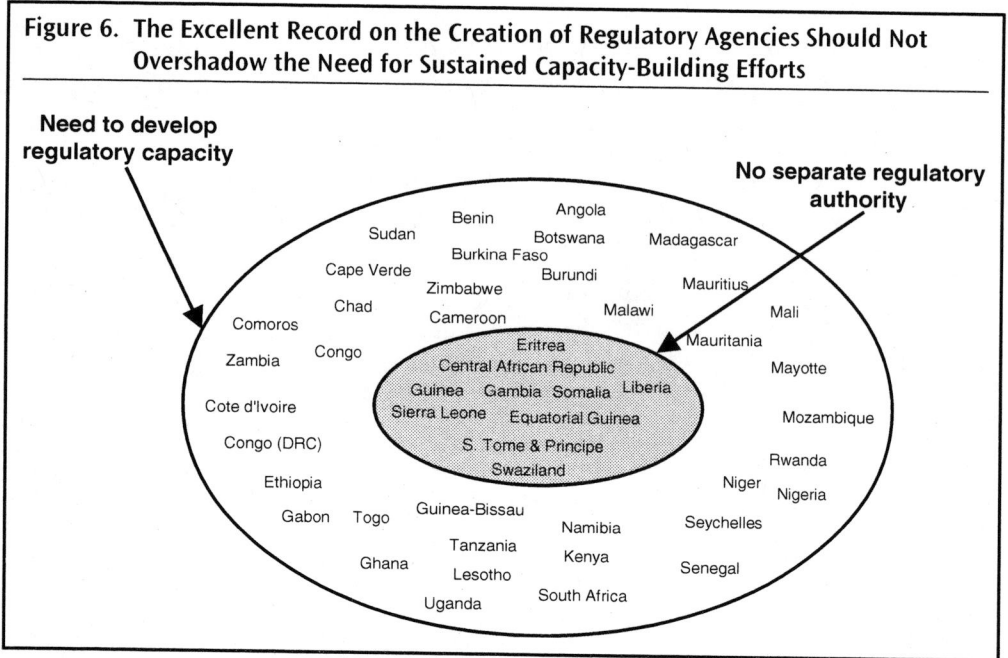

Figure 6. The Excellent Record on the Creation of Regulatory Agencies Should Not Overshadow the Need for Sustained Capacity-Building Efforts

Source: WBG GICT analysis, 2004.

by demonstrating a clear political commitment to put in place effective regulatory frameworks and institutions. Key areas where additional expertise and implementation of best practices would improve the certainty and predictability of regulation include: (a) dispute resolution mechanisms; (b) tariff control and tariff setting (at the retail and wholesale level) for interconnection; (c) leased line services; (d) licensing regimes; and (e) frequency management and compliance enforcement, particularly of the main incumbent or dominant players.

The vast majority of SSA countries have telecommunications regulators though some countries have yet to establish a regulatory body. The WBG is expanding its involvement in regulatory framework development and building the capacity of regulators to foster and safeguard competition for the benefit of current and potential users. In almost every country in SSA increased donor involvement is required to support capacity-building efforts at autonomous regulatory agencies.

Supporting Telecommunications Ministries

By far the most important role of SSA governments in the telecommunications sector is to ensure competition and level the playing field among key market players. At the most senior levels, the government must play a role in setting and helping to meet telecommunications access goals in a manner that supports private telecommunications operators. Ministries of telecommunications should set liberalization policies, which need to be supported by an adequately funded autonomous regulatory agency that can implement pro-competition policies. SSA ministries should then monitor progress, and develop indicators. In many cases, however, ministries of telecommunications in SSA lack the expertise and capacity to effectively shape the introduction of competition and sector growth.

As a first step, telecommunications ministries must draft and update ICT sector policy in consultation with the regulatory agency, market players, and other relevant stakeholders. As part of this function it is important that the ministry, in conjunction with the regulator, establish mechanisms to ensure that policies and regulations are not partial towards the public operators. Given that ministries in charge of telecommunications are often the major shareholder of public operators, it is critical that public operators be licensed in the same manner as private operators. Monitoring of license conditions should be placed under the oversight of the regulator. Together with the regulator and other stakeholders, the ministry in charge of communications should lay out a time bound transition to a fully competitive telecommunications regime.

SSA ministries should also be involved in aggregating demand for telecommunications infrastructure by different elements of the public sector in order to reduce the cost to government as a whole. Beyond these activities, the ministry should play a key role in promoting the harmonization of telecommunications policies and regulatory frameworks within regional, and subregional economic integration arrangements.[3] In Africa harmonization benefits investors because it increases the certainty of the regulatory environment across countries in SSA, as well as has the potential to accelerate reform in slower-moving countries in the region.

Telecommunications ministries in SSA need to build the capacity required to spur competition, and to monitor the progress of market development. The World Bank continues to increase its involvement in building the capacity of ministries of telecommunications to develop indicators, monitor market progress, and participate in regional harmonization efforts. Additional support by other donors is required to build more effective ministries in SSA.

Promoting Competition Through Pragmatic Liberalization

Market liberalization involves (among other things) opening the telecommunications sector to competition through the granting of licenses to fixed or mobile phone operators, ISPs, and broadband service providers. Liberalization strategies involve setting technical standards, defining price regulations, formulating licenses, designing interconnection principles and dispute resolution processes. *Pragmatic liberalization* in SSA requires the identification of key bottlenecks to sector development and creation of tailored actions to address them. While there is consensus in favor of the generalization of competition, its implementation requires the development of liberalization strategies that are customized to countries' characteristics.

Liberalization efforts need to move beyond the provision of mobile services. The case of mobile has demonstrated the extent to which the unleashing of competitive forces can unlock sector growth, can promote wider access to services, and can deliver tremendous benefits for countries. There is no doubt that diffusing competition across the sector will yield positive results. In some market segments (such as the provision of data services),

---

3. Such as the ECOWAS or the Southern Africa Development Community (SADC) in Africa.

incumbents stifle liberalized competition. In most countries ISPs are actually resellers of traffic and are unable to compete on an equal footing with incumbent operators, as the latter control access to infrastructure. ISPs have neither the right to build their own infrastructure nor the right to own their own international gateway. The lack of effective competition is not compensated for by adequate regulation. Thus, the lack of competition at the infrastructure level induced by the pattern of liberalization has not allowed competition to produce positive results.

As exclusivity periods for licenses granted to incumbent telecommunications operators are due to expire in the near future in many SSA countries, opportunities will be available to increase competition. An essential element of the reform of entry conditions is revising, updating, and strengthening legal and regulatory frameworks in order to accommodate multiple market players. Areas that deserve specific attention include dispute resolution mechanisms and interconnection, so that market players can respond to demand and, together, reach solutions. Demand for policy advice from governments is expected to increase.

Pragmatic liberalization strategies will be decisive in transitioning towards full liberalization given market and financial conditions in Africa. While the model of "second national operator" adopted in some countries has been successful in the past, further exploration is needed for innovative approaches that build on existing market players and suit the needs of individual countries.

The World Bank has identified 11 African countries where there is no competition to date in the area of mobile or international telephony services. The mobile telephony market, in particular, has been central to unlocking the growth of telecommunications service in Africa. As SSA countries express their interest in expanding competition across the sector, the WBG and other donors are willing to provide additional assistance in developing liberalization strategies. The WBG will also continue to assess the ways in which pragmatic liberalization can be used to expand access in SSA.

A Pragmatic Approach is Needed to Privatize Incumbents

Over the past two years, there has been a dramatic slowdown in the pace of telecommunications transactions in SSA. Negative investor sentiment has been exacerbated by a global equity contraction in the telecommunications sector. Easier privatizations have been carried out, but the privatization agenda has yet to be completed. Within the region, 17 countries have not initiated the transformation through privatization or corporate reform of their incumbent operator. Another 10 countries are in the process of doing so, but in 5 of them the privatization process is either stalled or is delayed considerably due to the lack of political commitment.

The fact that most incumbents in SSA are only partially privatized has far-reaching implications from a regulatory and competitive stand point. African governments face conflicting incentives, as they are the major shareholder of the incumbent. They can either ensure streams of dividends in the short-term, or empower fully functioning regulatory agencies to promote overall sector performance. In these cases, incumbent operators, which historically enjoy close relationships with governments, often win over newly established regulators.

In terms of structuring future privatization transactions, it is worth noting that half of the privatizations that have involved the sale of minority stakes are producing mixed

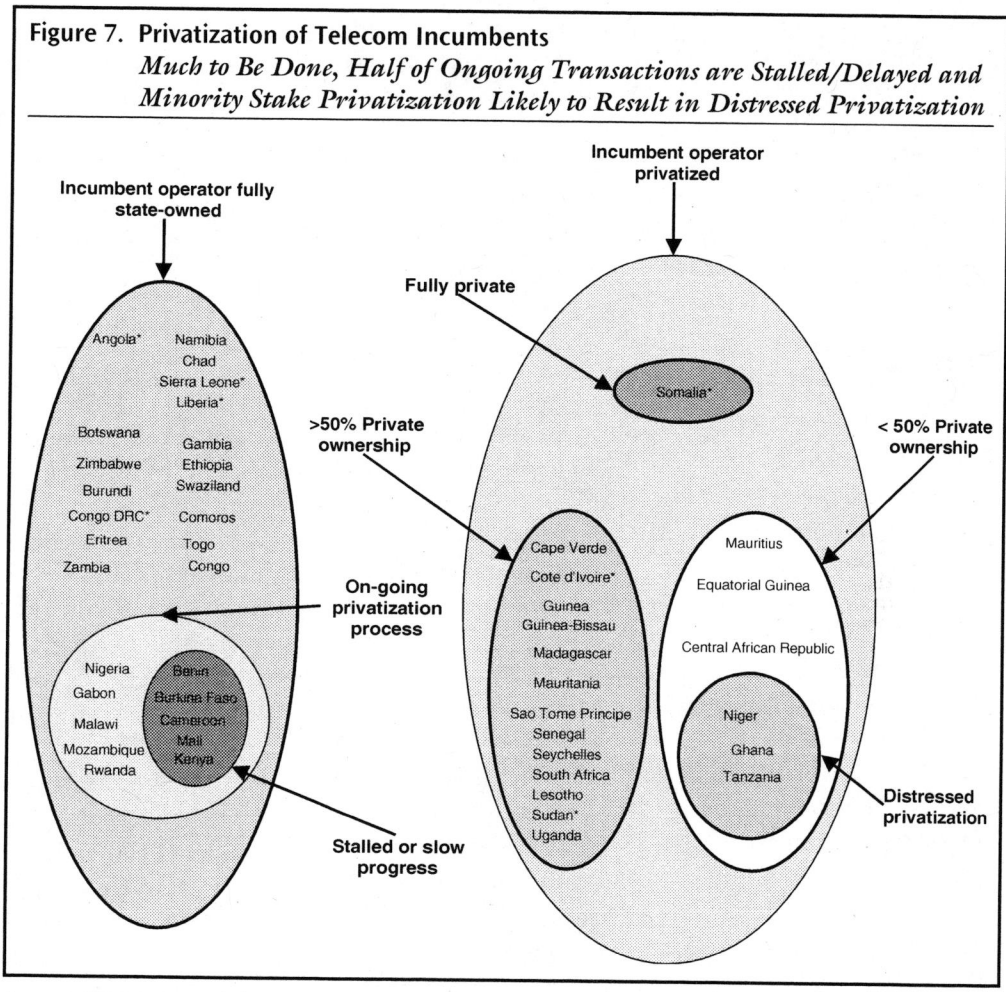

Figure 7. Privatization of Telecom Incumbents
Much to Be Done, Half of Ongoing Transactions are Stalled/Delayed and Minority Stake Privatization Likely to Result in Distressed Privatization

Source: WBG GICT analysis, 2004.

results and often lead to distressed relationships between the government and the private investors. Minority stakes often indicate a mixed signal from the government about their commitment to handing over telecommunications operations to the private sector. An important lesson can be drawn from Mauritania's experience with a transaction in which the government agreed to combine a sale of equity with a capital increase, which reflected a realistic and pragmatic approach to privatization. The privatization itself may not have maximized the proceeds to the government in the short term, but network rollout was accelerated to the benefit of Mauritania's citizens.

Access to finance will be improved if SSA countries have realistic expectations about privatization and license sale proceeds, and take into account the negative effect on investor sentiment caused by the delay of the privatization process or the failure to close on transactions. SSA governments can signal a commitment to transferring telecommunications operations to the private sector by packaging majority stakes, combining the sale of existing equity with a capital increase in the case of privatization transactions, and favoring investment commitments over transaction proceeds in the case of the sale of new licenses.

These methods of structuring transactions can significantly enhance Africa's attractiveness to investors. Conversely, failed or distressed incumbents inhibit sector development, discourage foreign direct investment, and provide negative signals to privatized incumbents seeking to initiate the privatization process.

The WBG continues to support countries in the privatization process by advising governments on appropriate strategies for increasing private sector investment in SSA telecommunications incumbents. The WBG will also support pragmatic approaches to increase private sector participation where outright privatization may not be an option.

Reforming the Postal Sector

Telecommunications sector reform creates opportunities for postal administrations to reorient themselves, modernize, and to offer ICT enabled services. The average mail volume in SSA is extremely low—below one letter per capita per year—which means that most public postal operators in SSA countries have negative net revenues, and drain state budgets. Where the postal sector receives government subventions/subsidies, there is a critical need to provide incentive to minimize cost and to put in place mechanisms of accountability and monitoring indicators. With the separation of telecommunications and postal functions—an integral component of telecommunications sector—it is crucial that SSA postal systems modernize and adopt corporate approaches to meet universal service obligations.

Postal sector reform in SSA can involve drafting new postal laws that introduce market liberalization strategies into the delivery of postal services. Legal changes can also result in the establishment of an independent postal regulatory body, which can help to ensure that universal service obligations are met, and service quality is maintained. Additional reforms involve modernizing postal enterprises. Possible strategies to make postal sectors more efficient and increase mail volumes are private-sector participation, commercialization of selected postal services, and corporatizing postal administrations (transforming them into publicly-owned companies governed by company law).

Many public postal operators manage postal financial services, while others usually distribute financial services through their network in partnership with a commercial bank. Some operators have installed Internet access in their main post office, and all of them contemplate investment in IT systems to extend Internet access through the postal network. Further targeted investments in IT and IT-enabled financial transactions—which can help turn post offices into transactional points for e-commerce—will be essential once postal sectors undergo reform and can effectively perform basic functions. For instance, Tanzania Posts Corporation has an ISP license and offers Internet services in six of its main post offices through a very small aperture terminal (VSAT) network. However, services are limited by the low bandwidth capacity and the lack of training and IT personnel.

All of the 16 African countries where the World Bank has recently been involved in postal sector reform have corporatized their public postal organization. However, of these 16 countries, only 4 have a regulatory agency that supervises the postal sector and that is separate from the line ministry and the public operator. The World Bank will continue to assist SSA countries with postal sector reform, and to encourage the donor community to support the pragmatic development of postal sectors in modernizing and incorporating ICT.

CHAPTER 4

Addressing Market Failures

In order for the SSA region to reap fully ICT's benefits, outstanding infrastructure gaps must be addressed. Even where the core reform agenda has been completed, specific public support may still be necessary to fill the critical gaps that markets cannot fill, such as: inadequate national backbone development, a lack of infrastructure and access in rural areas, and the poor quality of infrastructure in post-conflict countries.

National backbones are core infrastructure, and broadband backbones such as fiber-optic cable are needed to bring broadband Internet connectivity. In SSA greater access to broadband connectivity is required to provide governments, private-sector enterprises, and citizens with low-cost, high-quality bandwidth required to connect with the rest of the world. Given that digital information transmitted over backbone networks can be voice or data, national backbones also support the expansion of telephone, and Internet dial-up access through the interconnection of fixed and mobile telephone networks. While local loop or last mile issues pose challenges, clear deficiencies in the core transmission backbones prevent additional, more widespread infrastructure investment in rural areas, which is critical to reaching universal coverage in SSA countries.

Fostering ICT access in rural areas is important given that rural areas account for approximately 70 percent of SSA's population. Rural access enhances rural economic development and the functioning of rural markets. Rural telephone and Internet access create income-generating opportunities and can contribute to slowing the pace of migration to already over-crowded urban areas. Rural access contributes to the overall social and economic transformation of SSA countries. However, expanding ICT beyond urban areas presents major challenges. Specific policy interventions are required to attract private investment and to accelerate infrastructure deployment to low-density areas.

The dire lack of infrastructure in post-conflict countries that have been ravaged by civil or international war means that vast populations are entirely without access to ICT. Given that private sectors tend to be weak in these countries, that economies are often cash-based, that infrastructure is often severely damaged and, in light of governments' limited capacity to address these issues, donors and multilateral organizations must pay particular attention to these countries. ICT access can help rebuild war-torn economies and can facilitate reconciliation between former adversaries.

Developing national backbones, expanding rural access, and building infrastructure in post-conflict countries are costly and complex endeavors. Governments and the private sector will need to come together to create the infrastructure that is needed to expand networks. For their part, governments need to create incentives for private sector investment in national backbones, rural infrastructure, and post-conflict countries. "Smart subsidy" schemes involving auctions of subsidies with clear and specific teledensity targets may be efficient tools to encourage private-sector-led infrastructure development, and investment in rural areas—critical to spreading the benefits of ICT more equitably. However, policy and regulatory interventions should be the first tools used to address these issues.

The WBG will continue to support SSA governments in these areas provided that there is a proven market failure, that is, where competitive forces have been unable to achieve desirable policy objectives. However, it is critical that public support does not distort competition in nascent ICT sectors. The WBG will assist SSA governments to define policy and regulatory interventions that can foster infrastructure development.

Building National Backbones and Supporting Broadband Access

National backbone networks that support broadband access are typically long-term investments with significant sunk costs, high externalities, and public-good characteristics. All operators, including mobile operators, benefit from the existence of backbones because they facilitate cost-effective interconnection between the main city and secondary or tertiary towns. Such infrastructure allows for the provision of basic telecommunications services, but also allows for advanced value-added services, such as e-government, e-commerce, high-speed Internet access, and other economic development initiatives aimed at helping ICT to foster economic development and to reduce poverty. However, fixed backbone transmission and broadband networks have not been able to attract adequate levels of private investment.

The long-term rate-of-return profile of these projects makes them less attractive to private operators, which means they often require significant public support. Public support should first take the form of policy and regulatory interventions to spur private sector development. In the cases in which this fails, a possible policy option to consider is the development of OBA schemes (Box 2), though this type of support can only be seriously envisaged under specific and restrictive circumstances. OBA schemes could be used when there is a market failure in the form of limited or nonexistent backbone development despite a pro-competitive market entry regime allowing for the rollout and resale of infrastructure. OBA may also be an effective strategy where it is proven that sole private sector participation is not enough to secure the required financing.

Box 2. Public-Private Partnerships Can be Structured to Build National Backbone Networks

---

Significant evidence suggests that, in countries that improved backbone and broadband infrastructure, the government played a key role in spurring the private sector to provide increased broadband access and lower prices, and worked to ensure open access to broadband switches by ISPs and competing carriers by:

♦ putting in place and enforcing nondiscrimination requirements, and

♦ encouraging, in some cases, private operators to build a "public" national backbone through competitive OBA subsidy and tender mechanisms.

Support in this area by the development community could take the form of public-private initiatives, leveraging private sector investment, and accelerating infrastructure deployment through a new version of OBA-type schemes, which already is being effectively employed for provision of rural access services.

A number of African governments have approached the WBG for support in building national backbone networks, government intranets, or application-specific networks, often financed under WBG operations. Government intranet—or private networks that connect government agencies at the national, municipal, provincial, and local levels—could serve as the basis of a broader national network. However, the establishment of dedicated networks poses significant policy, economic, and financial issues that require careful consideration. All strategies for addressing internal access issues should be reviewed given that national backbone development is also crucial to increasing rural access.

Country-specific analytical work is being conducted by the WBG: (a) to assess existing supply and demand for backbone and broadband infrastructure; (b) to carry out a cost-benefit analysis of the options to be considered to develop this type of infrastructure; (c) to assess whether public funding is, in fact, needed to leverage private investment; and (d) to draw on the regional guidelines and provide practical recommendations for their implementation in SSA countries.

Designing Rural Access Strategies

Since telecommunications sector reforms began in the mid-1990s, there has been a dramatic increase in the number of people accessing phones. However, most of these users are concentrated in the urban areas where operators are willing to build their networks and investments make immediate and sound business-sense. This situation has left most of the semi-urban and rural areas off of the priority list, and without access to telecommunications services. With only five telephone (fixed or mobile) lines per 100 inhabitants in SSA, it is clear that a majority of semi-urban areas and rural areas are almost entirely without access.

Recent econometric analysis of the impact of telecommunications rollout (based on limited data) by Forestier, Grace, and Kenny (2002), indicates that the rollout of telecommunications infrastructure has actually encouraged the divergence in incomes within countries by targeting the wealthy in urban areas, whereas it could have played an important role in generating income and improving the well-being of poor and rural populations.[4]

Given that African economies (to a large extent) remain crop-based, and depend on farmers' productivity and output, enhancing access to information in rural communities would unleash untapped opportunities, by improving rural market operations and by transforming social lives in rural communities. ICT can increase the integration of the rural areas into the national economy through the improved:

- access to up-to-date and decentralized information on price,
- weather forecasts,
- opportunities for domestic markets, and
- opportunities for exports.

Many SSA countries have considered universal access schemes to roll out coverage to areas that are less attractive for investment, but a lack of funding has prevented these schemes from taking off. Universal service provisions, which are derived from European universal service regimes, typically require the local telecommunications industry to fund universal access. Often a percentage of revenue is channeled to a universal service fund that is established and managed by the independent regulator. While there are success stories (most notably in the Latin America region) in the use of smart subsidy models to promote rural connectivity, it can be challenging in countries where regulatory agencies with limited capacity would be further pressed by the demands of managing a universal access fund. A few SSA countries have started successful implementation of OBA schemes, as illustrated with the Uganda fund supported by the WBG.

Universal access mechanisms, subsidy schemes, and strategies for enhancing rural coverage obligations through the issuance of new licenses are promising vehicles to address rural isolation. In addition, emphasis should be placed on removing existing artificial barriers that prevent market players from creating solutions. For instance, depending on the design of the licensing regime, mobile operators could be allowed to install fixed public telephones using wireless technology, which could then be operated by private individuals. This solution is particularly relevant in semi-urban areas, and the use of signal boosters can extend the coverage of the existing network. Beyond the issue of subsidies, it is often largely a matter of adopting a pragmatic approach to regulating the sector.

The WBG is currently involved in rural access projects in a dozen African countries, mostly aimed at promoting private sector investment in deployment of services to rural areas through OBA schemes. In all cases these operations are part of a larger multisector reform program. Until now, most of these projects provided only sufficient funding for relatively limited pilot demonstration projects that are primarily aimed at rural telephony, and are typically targeted at a few villages in a region. The WBG continues to expand involvement in rural access schemes, and to encourage donor support for regulatory and policy interventions, as well as specific scaled-up financing interventions in rural areas.

---

4. Forestier, Emmanuel, Jeremy Grace, and Charles Kenny, "Can information and communications technologies be pro-poor?" Telecommunications Policy 26 (2002), p. 639.

Box 3. The OBA Approach to Universal Access Funds Disbursements in Uganda

Uganda faces several challenges in achieving universal access, with its per capita income of about $300 and a rural population of over 80 percent. Uganda began introducing sector reforms in 1996 and is reputed to have achieved one of the most competitive markets in SSA. In 2001, the Rural Communications Development Fund was created under the supervision of the Uganda Communications Commission, collecting annual contributions from all sector players in the amount of 1 percent of direct retail service revenues. Since 2003, the Rural Communications Development Fund has launched and successfully implemented small pilot projects involving the private provision of about 70 public telephones in underserved rural locations, 20 Internet points-of-presence in district capitals, 22 Internet cafés, 33 ICT training centers, and 26 district information portals.

The WBG is providing technical assistance to define nationwide projects for public telephony, Internet points-of-presence, and telecenters, as well as regulatory instruments, institutional arrangements, and bidding documents. The WBG is providing up to $10 million in a capital subsidy, which will finance over 80 percent of the subsidy requirement for these projects. During 2005 implementation will be launched for two projects: a public telephony project, involving private provision of over 800 public telephones in underserved rural areas; and a project involving the provision of 32 Internet points-of-presence in district capitals.

Source: WBG GICT Analysis, 2004.

Supporting Countries in Their Post-Conflict Activities

Restoring some level of ICT infrastructure in post-conflict environments remains a major priority. A widespread broadcast footprint can broadly disseminate information on the location of needed emergency relief, and can be put in place relatively quickly. Basic telephone capacity in major cities also allows for government communications and donor coordination. By enabling communication, ICT also helps to reduce the chances of conflicts, and increases the opportunities to overcome conflict. In post-conflict countries, restoring a minimal level of ICT infrastructure is generally a top priority to facilitate emergency relief coordination and reconstruction efforts, as well as to restart the economy.

Previous experience shows that policy reform technical assistance in post-conflict countries usually leads to considerable private investment. In Afghanistan, for example, competitive private investment in mobile networks constructed since 2002 already approaches $100 million. Transparent licensing of new operators can provide the needed confidence to potential local and international investors to enter the difficult markets. In Liberia, for example, the World Bank is assisting the government to rationalize its licensing and spectrum policies as a first step towards more comprehensive sector restructuring. Initial financing of critical parts of the backbone infrastructure can facilitate the delivery of some government services, while improving the probability and pace of the rollout of privately financed and operated networks. Development of other social sectors (like education and health) can be accelerated through an initial boost to the ICT sector. Not to mention that ICT development inevitably creates employment opportunities that spill over to other infrastructure industries.

The WBG is reviewing ongoing post-conflict ICT interventions in order to create strategies for resolving ICT constraints and identifying effective policies and financial instruments to be applied in post conflict countries. Possible interventions may include:

▓ the design and provision of initial financing for emergency ICT network restoration projects, and
▓ assistance in mobilizing donors in financing and co-financing of such emergency investment projects.

There are additional opportunities for joint approaches with the International Finance Corporation in designing pre-privatization investment projects and facilitating foreign direct investment.

The telecommunications sector can attract significant private investment, even in the immediate aftermath of conflict and during periods of conflict—as seen in the Democratic Republic of Congo, Liberia, Sierra Leone, and Somalia—important evidence of the benefits of private-sector led development. In the Democratic Republic of Congo, the private sector has invested over $380 million in telecommunications projects in 2003 and in the first three quarters of 2004. To grasp the importance of this aggregate telecommunications-related investment, it is interesting to compare this figure with the $700 million of WBG support to that country in FY04. However, the lack of regulations common to post-conflict environments increases the risk that these investments be made into businesses that will not transfer their value as they should. This clearly shows the urgency of policy and regulatory support to allow immediate competitive private-entry into the sector.

The WBG is currently assessing strategies for further involvement in SSA post-conflict situations. The unique complexities of the post-conflict environment require a two-track response from the development community. First, donor support is needed for the urgent construction and reconstruction of networks to serve vital government communications needs at the national and regional levels, including broadcast communication, as well as basic government telecommunications networks. Second, and in parallel, donors need to provide significant technical support to allow private competitive investment to take an early and sustainable role in sector development.

CHAPTER 5

Promoting "ICT for Development" Applications

Many SSA governments have developed (or are in the process of developing) national ICT strategies, which bring together government ministries, private sector enterprises, and civil society organizations in an effort to create a holistic approach to ICT sector development. While there are constraints to creating any form of national strategy, particularly given the limited capacity and resources available in SSA, the main principles of national ICT strategies are to remove the policy and institutional barriers that prevent governments, civil society organizations, and private sector enterprises from applying ICT to meet development objectives. Currently, a primary barrier to applying ICT remains the lack of infrastructure in SSA. As a result, any progress made on infrastructure development through national ICT strategies is important.

Whether they are the result of a national ICT strategy, "ICT for development" applications support specific development goals, and are critical to leveraging the enabling capacity of ICT. "ICT for development" applications can range in focus; for instance, e-commerce applications can support increased interregional trade, local private sector development, and rural agricultural market development. Civil society applications, ranging from e-education to e-health, can help to achieve social objectives such as healthcare delivery and HIV/AIDS prevention. E-government applications help to achieve institutional objectives such as increased efficiency in public service delivery and provision, government accountability, and transparency. While "ICT for development" applications naturally require infrastructure and access, increased investment in application development may, in turn, help to spur investment in infrastructure and expand access to rural areas. Targeting applications is a key part of developing a robust ICT sector in SSA.

The WBG is developing measures to monitor and evaluate national ICT strategies in a small group of countries. The main emphasis is on developing common sense approaches to designing ICT strategies whose implementation can be monitored and assessed. Increase in donor involvement in national ICT strategy development is needed given that these strategies can trigger important dialogue between the private-sector, governments, and civil society organizations on priorities for ICT sector development at the national level.

Facilitating E-Commerce

Promoting e-commerce and other ICT applications is the linchpin of the "new economy." Global communication networks increase a country's trade prospects, open new markets, and promote development by enabling leading primary, industrial, and services sectors—as well as small-scale enterprises—to trade with larger markets. E-commerce applications actively support trade diversification and export promotion in SSA, as well as the removal of bottlenecks to intraregional trade. E-commerce transactions over the Internet, however, open a host of difficult legal issues.

Some of the key policy and legal issues that need to be addressed in order to create a viable business environment on the Internet in SSA, include:

- the recognition and enforceability of *electronic contracts,* provisions to ensure that *digital signatures* can be authenticated and are legally binding, and retention of electronic data;
- the development of an information security architecture, certification processes, and regulations to ensure confidentiality, and *security* of transactions and financial data;
- ensuring that necessary *encryption* technologies (including the underlying software) can be imported and deployed to provide the security and certainty required to encourage users' confidence to conduct electronic commerce;
- legal provisions for *protecting the privacy of personal data;*
- world-class protection of *copyright and intellectual property rights* for electronic data; and
- the reduction of import barriers, *duties and import tariffs,* to high-tech hardware and software necessary for e-commerce to flourish.

While e-commerce markets remain small in SSA due to severe infrastructure constraints, and existing limited legal and regulatory capacity to ensure the safety of transactions and personal data exchange, countries in the planning stages should identify ways to nurture business-to-government, as well as business-to-business services. In particular, countries should encourage services that relate to government procurement, international trade, and IT and Web services. SSA governments can take the lead in conducting international and national e-commerce transactions, increasing both transparency and governance efficiency, while also helping to create demand for the infrastructure that can support e-commerce.

The WBG is exploring ways to facilitate e-commerce development in SSA. Donors, the private sector, and civil society should review the regulatory and policy impediments to creating e-commerce services, particularly as a mechanism for further supporting infrastructure development and expanded access.

Supporting E-Government Applications

In SSA, government applications of ICT in the form of e-government, e-procurement, and e-administration services have already contributed to the emergence of good governance practices and accountability. ICT incorporated into government processes has streamlined procedures and operations, and has reduced bureaucracy, which, in turn, assists in the effective delivery of public services.

A wider diffusion of ICT applications and the increased use of ICT services by governments—resulting from the development of government networks and appropriate regulatory frameworks—can further enhance the capacity of governments to deliver services more efficiently, to promote effective decentralization, and to improve national systems of accountability and governance structures. By implementing e-administration and procurement applications, governments can also help to spur the development of nascent e-commerce industries.

There is a growing recognition that governments should develop and implement a comprehensive and integrated national ICT strategy in order to avoid duplication in infrastructure of private network development and ensure maximum effectiveness of e-government. Such a strategy needs to address both the deployment of infrastructure from economic, financial, and regulatory angles, as well as to ensure the effective use of the infrastructure through e-government applications and networks. In addition, even without the infrastructure component, deployment of advanced ICT applications in environments where the infrastructure is still monopolized by one private or public sector operator is often unsustainable due to high prices and low quality of service.

Beyond decentralizing government services, ICT has the power to help marginalized citizens participate in their nation's policies. ICT can act as a tool for inclusion to connect governments to their citizens, citizens to their government, and citizens to citizens, as has occurred in South Africa (Box 4).

The WBG is exploring ways to support the development of e-government strategies, particularly as a part of national ICT strategies. Donors and governments should conduct more analytical work to identify international best practices for e-government application development, and effective means by which to monitor and evaluate such types of programs.

Box 4. South Africa: An Example of a Government Using ICT for Opened Decision-making: The Drafting of Post-Apartheid Constitution

In South Africa, the government sought citizen input to draft the post-Apartheid constitution by creating a Web site that made available all related documents (draft texts, political party positions, committee reports and recommendations). In addition, citizens posted their feedback online, were given the opportunity to send emails to the government, and to share ideas on bulletin boards (Lal 1999).

Source: Grace Jeremy, Charles Kenny, and Christine Zhen-Wei Qiang, *Information and Communication Technologies and Broad-Based Development, A Partial Review of the Evidence,* World Bank Working Paper No. 12, December 2003.

Fostering Civil Society Applications

The increased use of "ICT for development" applications in health and education supports the development of human capital and the strengthening of educational and health systems as well as the empowerment of disadvantaged groups of citizens. A number of civil society organizations, in SSA and around the world, are involved in promoting ICT as a mechanism for achieving development objectives in the region. However, the success of these programs is largely conditional upon the availability of information and communication infrastructure. Smaller pilot initiatives that provide Internet access to schools, or hospitals may only have a limited impact, particularly since many of these initiatives lack sustainable funding and support.

ICT-based education applications can support learning and provide critical curriculum and teacher support. Distance learning courses, CD-ROM educational materials, and online collaborative learning spaces can support student and teacher training, and supplement education in SSA countries. Box 5 showcases two examples of successful e-learning initiatives.

Internet access also opens doors for isolated rural, severely disadvantaged or handicapped students. As illustrated with the African Virtual University (AVU), the use of ICT in primary, secondary, and university education can support the development of technology-literate population, a necessity for the global knowledge economy.

In addition to e-learning, health e-applications can have a powerful impact in many parts of SSA, particularly in rural areas. Together, geographic information systems, the Internet, and telecommunications can assist to coordinate health care delivery in areas where diseases runs rampant. In many cases, ICT could be used to alert health care practitioners of disease outbreaks, to help expeditiously isolate and limit the spread of the disease (Box 6). While concerted efforts need to be made to overcome infrastructure challenges in SSA, telemedicine techniques could be applied to reach isolated medical clinics, providing them with up-to-date medical information and connecting them with centralized health ministries. Civil society organizations, governments, private sector health care providers, and donors are currently experimenting with several approaches in SSA involving satellite Internet access. National ICT strategies should address health components in a more comprehensive way if these benefits are to be extended in a more holistic manner.

The WBG is developing mechanisms for monitoring and evaluating national ICT strategies, and civil society applications are often a key component of these strategies. More analytical work needs to be done to create the appropriate models for public funding of e-health or e-education initiatives. Donors and the development community should support the development of policy and regulatory frameworks, as well as infrastructure that can support e-health, and e-education applications in SSA.

Box 5. The Impact of ICT on Learning

World Bank Institute: The Global Development Learning Network (**GDLN**) was launched in June 2000 by the World Bank and several partner institutions to develop e-learning platforms where ideas and experiences could be shared between policy makers and development practitioners in a cost-effective way. GDLN has a global network of 60 affiliated centers, and every year connects more than 30,000 people. GDLN offers e-workshops throughout the year and brings participants together within and across regions.

From GDLN affiliated centers, in October 2003, participants in Ethiopia, Tanzania, and Uganda exchanged experiences in a series of workshops on gender, and the legal dimension of HIV/AIDS. The objectives of the series were to strategize on strengthening the role of laws and institutions in HIV/AIDS policy and operations.

Source: ICT and MDGs, A WBG Perspective, December 2003.

The *African Virtual University* (**AVU**) is a World Bank project initiated in 1997, with the objective of gathering global knowledge to meet Africa's educational needs. Today AVU operates 34 learning centers in 17 SSA countries and has provided courses to 23,000 participants, from language and Web design to women entrepreneurship.

The World Bank has played an essential role in forming strategic partnerships with the private sector to set up the infrastructure required to broadcast AVU courses. The Bank has rallied the international development community to the project: Australia, Belgium, Canada, Ireland, Norway, Sweden, the United Kingdom, the United States, and the European Union have joined the Bank in providing technical and financial support to AVU. By 1999, 30 major institutions around the world were involved in the virtual learning network. The focus of AVU has shifted to building the capacity of local African institutions to generate their own courses, therefore promoting South-South and South-North cooperation. The lessons learned from this successful venture are the following:

(a)   Commitment from local governments and universities made the success of AVU possible.

(b)   Technology is a critical factor of success in establishing e-networks. Strategic partnerships with the local/international private sector provided both the IT equipment and the technical expertise to run AVU. The training of local staff allowed the project to run and, more generally, contributed to the development of Africa's human capital.

(c)   The sustainability of the project is conditional upon local fund-raising. The pilot AVU project relied too heavily on donor financing and private sector subsidies. AVU headquarters were relocated from Washington, D.C., to Nairobi, Kenya where it now operates as an independent nongovernmental organization, with self-financing learning centers across SSA.

Source: Analysis based on Prakash Siddhartha, "The African Virtual University and Growth in Africa," World Bank, February 2003.

Box 6. Some Applications of ICT on Health Care

In Uganda 54 percent of health care workers, and in Kenya 20 percent of health care workers have participated in radio training courses that lead to improved health care services (Kenny, 2003).

In Gambia, health care workers in rural areas use the Internet to send images of a patient's symptoms to physicians for diagnosis. Similar pilot projects are on the rise throughout the region.

In West Africa, ICT has been used to control Onchocerciasis (river blindness); local inhabitants send data collected by sensors along rivers to entomologists who then calculate the optimum time to spray against the disease-carrying blackfly. As a result, river blindness has been eradicated in seven countries, enabling 30 million people to live free of the disease (Kenny, Navas-Sabater, and Qiang, 2002).

Source: ICT and MDGs, A World Bank Group Perspective, December 2003

CHAPTER 6

Accelerating Regional Integration and Connectivity

The lack of adequate regional infrastructure to support high-quality telecommunications in SSA is recognized as a major obstacle for setting the region's economic and social development in motion. Regional telecommunications is both limited and expensive for SSA. In addition, the region has the lowest capacity in the world for international Internet bandwidth. This lack of bandwidth, in turn, helps to prevent SSA from taking active part in the global economy.[5]

Increasing regional connections to global fiber optic cables will help to reduce the cost of accessing all forms of ICT. Instituting regional roaming between SSA countries would make business communication easier and less costly. Developing a regional Internet node, as well as national exchange points, would reduce the cost of transmitting Internet information in SSA. Lastly, harmonizing policies and regulations across SSA countries will do much to create a regional market.

The WBG maintains a regular policy dialogue with the key regional players, including the e-Africa Commission of NEPAD, the African Union Committee of Ministers in charge of Communications, the African Telecommunications Union, as well as Regional Economic Communities, such as ECOWAS, Southern African Development Community, and the Common Market for Eastern and Southern Africa. Most of these institutions require

---

5. According to the latest ITU survey, the Africa region has the lowest amount of international Internet bandwidth in the world with 1,236 megabits per second (Mbit/s) in 2001, less than one-eighth of Oceania's 8,968 Mbit/s and a miniscule fraction of the 353,040 Mbit/s for the Americas. Conversely, a recent ITU survey found that the Africa region also has the highest broadband prices of $913 per month compared to $320 per month for Oceania and $88 for the Americas ITU ("Birth of Broadband," September 2003, Tables A-47 and A-51). The prices are for monthly residential subscription cost to broadband service based on the most "common" or cost-efficient broadband offer. See technical notes in *Broadband for All* for details (page A-56).

additional support if they are to play their role effectively, driving sustainable development of regional infrastructure.

Promoting Regional Harmonization

Harmonization of policies can support cross-border connectivity in a variety of forms; with harmonized policy and regulatory approaches, issues such as fragmented regional markets and low intracommunity traffic volumes can be addressed. The increased regional traffic that harmonization can bring can also help spur the development of national backbones, which can then maximize the benefits of cross-border links. Given that several regional infrastructure plans and projects are being developed in SSA— including fiber optic projects between SSA countries, regional submarine cable networks, digital microwave transmission links, and VSAT networks—additional research first needs to be conducted to determine the structure of the market for cross-border connectivity in the region.

To advance a program of regional harmonization, more detailed assessments of regional and cross-border connectivity gaps and clear strategies for filling those gaps are needed. Quantitative analysis of the current demand for cross-border infrastructure will help to justify the importance of improved connectivity and help to create momentum for dealing with these gaps. African regional communities are undertaking projects to harmonize their telecommunications, such as the ECOWAS project (Box 7). Rather than generating new policies and regulations on their own, smaller SSA countries are increasingly using sub-regional organizations to find appropriate models.

Policy and regulatory changes will be required for the success of any initiative linking the subregion. Policy and regulatory frameworks that affect cross-border connectivity need to be addressed, and ways to overcome barriers to deployment of cross-border projects should be explored. The WBG is assisting with the development of assessments that will determine the market for cross-border connectivity, and appropriate strategies for advancing a program of regional harmonization.

Box 7. Working Towards the Establishment of an ECOWAS Common Telecommunication Market

As part of its effort to promote regional economic integration, ECOWAS, supported by the World Bank and Public Private Infrastructure Advisory Facility has launched a program to develop a common framework to facilitate the harmonization of the national telecommunication sector policies throughout member states with the ultimate goal of establishing a telecommunications common market within the ECOWAS region.

ECOWAS commissioned a telecommunication harmonization study (launched in February 2002), for harmonizing national legislative and regulatory arrangements with a view to evolving a telecommunications common market for the region. The study provides a set of recommendations on proceeding with the harmonization program with the view of reinforcing investment in the sector. Subsequent ministerial meetings in Abuja, Accra, and Lomé have validated the results of the study. A strategic vision statement and objectives for 2010, as well as short- and medium-term action plans have emerged from these meetings and are key to the harmonization process.

Increasing Regional Connectivity

African heads of state expressed their strong support for initiatives that will help create regional infrastructure in SSA and increase connectivity throughout the region; they recognize that regional integration of telecommunications makes private sector investment in infrastructure more attractive and leads to regional economic integration. A great majority of intra-African traffic is being routed through Europe and elsewhere via satellite links resulting in high transit charges and a significant financial drain. In response to this urgent need, NEPAD has highlighted international broadband connectivity as one of its six ICT priority projects.[6]

Developing Regional Fiber Optic Cable Projects

Africa currently pays in excess of $400 million annually in fees for calls transiting through Europe and the United States. The East Africa region is even more poorly served, being the only part of coastal Africa that does not have direct connections to global fiber optic networks. Instead, the region relies entirely on satellite connectivity, which is more expensive and of lower quality, thus hindering the region's potential for economic development.

A growing number of SSA countries, governments, regional organizations (notably NEPAD, SADC, Common Market for Eastern and Southern Africa, and ECOWAS), and private-sector operators have identified building regional backbone infrastructure as a top priority for improving connectivity in the region. Several regional backbone infrastructure initiatives are currently under discussion across the region, in particular the East Africa Submarine Cable System (Box 8), COMTEL, COM-7, the Southern African Development Community's Regional Information Infrastructure, and iFoni. The cost of each project ranges from $60 million to over $300 million.

Of paramount importance is the need to support NEPAD and other regional institutions to develop consensus on open access and a common policy and regulatory platform for regional integration. In addition, governments must be encouraged to accelerate capacity-building for national regulatory authorities, so that they can efficiently deal with competitive access to cross-border infrastructure, national infrastructure sharing, liberalization of international gateways, conducive interconnection regimes (and enforcement of those), and provision of regional licenses.

Beyond policy and regulatory considerations, it is also becoming increasingly apparent that alone, private sector capital may not be sufficient to fill this cross-border connectivity gap. The WBG, with other partners in the donor community, will explore options for contributing to the investment in cross-border infrastructure and national backbones for coastal and landlocked countries willing to ensure open and competitive access to regional infrastructure and/or transparent and nondiscriminatory infrastructure-sharing arrangements for national backbones.

The World Bank Group, through its Africa Regional Integration department and ICT Policy division, is currently assisting the NEPAD secretariat and several governments to examine

---

6. The Heads-of-State and Government Implementation Committee of NEPAD noted the six ICT priority projects in the March 9, 2003 communiqué (cf. Part I).

Box 8. Accelerating Connectivity in Eastern & Southern Africa (E&SA)

The World Bank and its development partners are supporting several projects aimed at linking E&SA countries to one another and to the rest of the world by 2010. To meet this objective, the following must be achieved:

♦ the gap must be closed the in optical submarine cable loop around Africa;

♦ all land-locked countries must be connected to submarine cable systems; and

♦ an integrated, continent-wide broadband ICT system must be established.

The East Africa Submarine Cable System (EASSy) and COMTEL are two of the projects designed to contribute to cross-border connectivity throughout Africa and recognized by NEPAD as two of 19 ICT priority projects:

♦ EASSy is a fiber optic cable project proposed to connect seven coastal countries in East Africa, including: Djibouti, Kenya, Madagascar, Mozambique, Somalia, South Africa and Tanzania (the "EASSy anchor countries"). In addition, more than a dozen other East African countries have expressed interest in connecting to EASSy through terrestrial means. With EASSy, submarine cables will surround the African continent for the first time. Both SAT-3 and the planned EASSy will be used for building additional terrestrial broadband networks that would connect the submarine cables to landlocked SSA countries.

♦ COMTEL is a regional terrestrial fiber optic cable backbone network which plans to connect all of the unconnected countries in East Africa with the exceptions of Somalia, Mauritius and the Seychelles. In the short-term it will be a regional network delivering traffic into SAT3 in South Africa and the long-term plan is to connect to an international fiber link, both in Djibouti and Egypt.

These projects will help Africa move to move towards self-sufficiency and decrease the continent's dependency on outside countries for telecommunications services. Instead of paying high charges for international connection through a transit point, operators in the region can establish direct connections, promising a substantive decrease in their operating costs for international telecommunications. Unverified reports indicate that the SAT-3 cable, a fiber optic cable that runs along the western coast of SSA, is saving members about $400 million a year because they are no longer forced to route their traffic through satellites belonging to the United States and Europe. However the closed club configuration of the SAT-3 cable means these savings have not necessarily been passed on to consumers.

In order to secure maximum benefits from these new initiatives to African consumers and businesses through substantial price reduction in voice and Internet services, restrictions and bottlenecks on international access need to be addressed. Implementing open access business models and empowering regulators are possible options. In the long-term, the benefits of improved connectivity and improved competition will result in an increase in the competitiveness and attractiveness of SSA to foreign investors, and accelerated economic and social development.

---

the legal and regulatory barriers in the SSA region, and facilitate consensus-building activities between SSA governments and regional organizations focused on prioritizing and rationalizing the competing initiatives. Regulatory and policy conditions of participating countries are being assessed, appraised, and adjusted to ensure open, fair, and pro-competitive access to the proposed infrastructure. The International Finance Corporation, in collaboration with the Development Bank of South Africa, has provided initial funding to undertake a feasibility study of the East Africa Cable. Beyond this support, the WBG, with other partners in the donor com-

munity, will explore options for contributing to the investment in cross-border infrastructure and national backbones for coastal and landlocked countries.

Supporting Roaming Initiatives

Regional voice traffic routed outside the region and transiting through third-party countries is extremely costly for telecommunications operators, the African consumer, and the entire SSA economy. While the mobile sector is increasingly recognized as holding a special potential for the African continent, the lack of adequate regional roaming is also a major obstacle for cross-border mobile connectivity, preventing mobile technology from fulfilling its full potential to serve pent-up demand not otherwise satisfied by the fixed networks.

Promoting mobile roaming across SSA requires identification of the policy and regulatory safeguards that create the enabling environment for the success of cross-border initiatives, such as licensing and authorization regimes, a regional licensing framework with common guidelines, and consistent license application procedures and time lines. Additional initiatives that will support regional roaming include enforcing transparency, providing sufficient information on roaming retail prices to end-users, identifying the cost and benefits of adopting a benchmarking approach for retail prices, and promoting competition in the mobile roaming market.

The WBG, in collaboration with the Public Private Investment Advisory Facility (PPIAF), is supporting an ongoing initiative to identify and remove bottlenecks to roaming, cross-border connectivity, and international gateway access in the ECOWAS region. The initiative will also help to identify priority cross-border links for which effective public-private partnerships and international donor financing could play a valuable role.

Building a Network of National Exchange Points

Additionally, Internet exchange points (IXPs) are highly effective mechanisms for reducing the cost of Internet connectivity and bandwidth, and improving quality of service. IXPs interconnect ISPs in regions and countries, enabling ISPs to send Internet traffic within countries. Without national IXPs, Internet traffic is routed across international exchange points at often very high costs.

Only nine of SSA's countries have national IXPs.[7] The rollout of the Kenyan national IXP (Box 9) has resulted in increased competition, higher quality of service, and a decrease in leased line prices. Nearly every African ISP must rely on satellite connectivity, which is more expensive than the use of fiber optic cable. The deployment of additional IXPs in SSA is inhibited by prohibitions on unregulated telecommunications facilities, regulatory agencies that seek to establish control over the Internet, and large telecommunications operators and ISPs that seek to prevent effective competition. In order to achieve wider IXP deployment in SSA, ISPs can organize into associations to administer facilities, and call for regulatory reform and liberalization.

---

7. Democratic Republic of Congo, Kenya, Mozambique, Nigeria, Rwanda, South Africa, Tanzania, Uganda, and Zimbabwe.

Box 9. The Impact of the Kenya IXP on Quality and Cost of Infrastructure Services

---

The Kenyan national Internet Exchange Point (KIXP) was established on February 14, 2002. Initially four ISPs were exchanging traffic. In August 2003, the network had increased to 10 ISPs, namely SwiftGlobal, Kenyaweb, ISPKenya, UUNET Kenya, Interconnect, Wananchi Online, AccessKenya, Nairobinet, Mitsuminet, and Insight Kenya.

A consultation process led by DFID found that the increase in the number of ISPs can result in: robust African backbones with exchange points at the core; savings on international expenses, robust low cost local infrastructure; and local environments conducive to e-commerce, e-finance, telemedicine, e-learning, streaming video and audio, etc.

In Kenya, latency for traffic exchange fell from 1,200 to 2,000 microseconds (ms) on average to 30 to 60 ms on a link to the Kenyan Internet Exchange Point. Similarly, leased lines prices in Kenya dropped almost fifteen-fold.

Quality of Service and Exchange of Domestic Internet Traffic

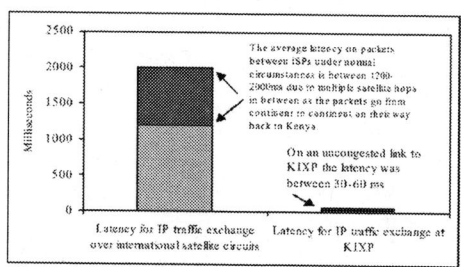

Difference between International and Domestic Leased Line Prices in Kenya, December 2000

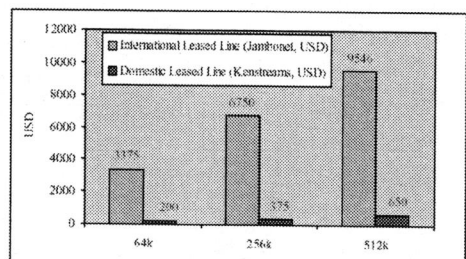

Source: Current and Future Status of National and Regional Exchange Points in Africa, NAPLA2003, Buenos Aires, Argentina August 2003.

In addition to providing robust and low cost infrastructure that is conducive to the expansion of ICT applications, the establishment of national IXPs is also critical to successfully roll out regional infrastructure backbone networks.

The development community is currently supporting the work of African ISPs to build IXPs in SSA through analytical studies and other capacity-building support. Donors are assessing the impact of the IXPs on national, regional, and continental traffic flows and the African Internet economy as a whole. It is expected that between 20 and 30 IXPs will be established over the next few years.

Internet exchange points are critical to the development of Internet infrastructure in SSA. Only with additional national IXPs in the region can Internet traffic be routed at a lower cost within SSA. Lower cost Internet access will lead to more use of the Internet by the private sector, civil society, and government, as well as expanded individual use. The World Bank encourages donors to increase their involvement in the development of IXPs in each SSA country.

Conclusions

There is little doubt among SSA countries that increased access to ICT infrastructure—the foundation upon which the ICT sector is formed—is a critical engine for growth and competitiveness, as well as a major contributor to poverty reduction. Africans have made infrastructure development and access a priority and have highlighted this at several national and international forums—most recently at the World Summit of the Information Society in Geneva (November 2003)—and continue to request support from their developing partners to invest in this critical area.

In truth, significant improvements have occurred in the ICT sector over the past decade, and the result is an unprecedented explosion of mobile communications in most of Africa. However, an assessment of the core ICT reform agenda in several countries (several of which have been supported by the WBG) indicate that despite these improvements, some key issues remain which impede sector growth. In a number of cases, pro-competitive legal and regulatory frameworks, required to create a competitive marketplace, are missing. Analyses of the lessons learned from former privatizations show, for example, that the privatizations that attract the most investments are those structured through a combined sale of equity with a capital increase, as they signal a strong political will to transfer operations to the private sector.

In general pragmatic liberalization strategies tailored to the context of each country, and backed by sound policy and regulatory frameworks, have the highest chance of success. The WBG has played a significant role in shaping earlier reforms and will continue to assist governments in developing more pragmatic strategies, as well as to hone the capacity of regulatory agencies and the telecommunications ministries to effectively manage sector development. In addition to the ongoing support of the reform agenda, the new challenge is to assist governments in implementing this second generation of pragmatic reforms. Equally challenging is the expanded agenda of infrastructure development needs, including the delivery of national and regional broadband connectivity to all countries (including those emerging from conflicts), the development of rural access, and the roll-out of "ICT for development" applications.

The WBG is working closely with donors, governments, the private sector, and civil society organizations within SSA to consolidate progress made in the telecommunications and ICT sector in the past decade. The expectation is that this will serve as a basis for a revitalized ICT agenda for the region, and that it can be matched with the necessary investment capital and resources. The second meeting of the World Summit on the Information Society in Tunisia (in November 2005) will be a critical opportunity to bring governments, donors, and the private sector together to assess the first concrete results of the renewed emphasis on the ICT sector in SSA.

ICT Indicators in
Sub-Saharan Africa

	Total Telephone Subscribers pr 100 inhab. (in%)								
	Sept 2003	2002	1995	CARG 95–02 (in %)	Sept 2003 Ranking	2002 Ranking	Change of Ranking (95–03)	Change of Ranking (1995–2002)	% of Mobile in Total (2002)
Angola	2.35	1.54	0.51	17	31	32	–6	–7	60
Benin	4.58	4.14	0.54	34	24	22	–2	0	78
Botswana	34.65	32.85	4.09	35	4	4	3	3	73
Burkina Faso	1.90	1.29	0.30	23	34	38	0	–4	58
Burundi	1.11	1.06	0.29	20	42	40	–6	–4	70
Cameroon	6.73	4.97	0.51	38	18	20	6	4	86
Cape Verde	27.38	25.77	5.57	24	5	5	–1	–1	38
Central African Rep.	0.87	0.55	0.25	12	46	46	–8	–8	58
Chad	0.87	0.58	0.08	32	45	45	3	3	74
Comoros	5.08	1.35	0.72	9	21	36	–2	–17	0
Congo	9.31	7.39	0.81	37	12	15	6	3	91
Congo (DRC)	1.65	1.08	0.10	41	38	39	9	8	98
Cote d'Ivoire	8.68	8.27	0.86	38	14	12	3	5	75
Equatorial Guinea	8.47	8.08	0.63	44	15	13	6	8	78
Eritrea	0.89	0.90	0.49	9	43	42	–17	–16	0
Ethiopia	0.61	0.60	0.25	13	48	44	–9	–5	12
Gabon	22.59	23.97	3.35	32	6	6	2	2	90
Gambia	12.09	10.08	1.89	27	9	10	2	1	72
Ghana	4.74	3.34	0.41	35	22	23	6	5	62
Guinea	1.62	1.52	0.17	37	39	34	3	8	78
Guinea-Bissau	0.88	0.89	0.69	4	44	43	–24	–23	0
Kenya	7.35	5.18	1.01	26	17	19	–4	–6	80
Lesotho	8.16	5.57	0.88	30	16	16	0	0	76
Liberia	1.34	0.28	0.16	8	41	48	3	–4	23
Madagascar	1.79	1.40	0.30	25	36	35	–1	0	73
Malawi	1.46	1.52	0.37	22	40	33	–8	–1	54
Mali	2.45	1.03	0.19	28	28	41	13	0	48
Mauritania	11.07	10.39	0.41	59	10	9	17	18	89
Mauritius	60.20	55.95	14.25	22	2	2	0	0	52
Mayotte	20.84	21.64	4.66	25	7	7	–1	–1	68
Mozambique	2.37	1.86	0.40	25	29	28	1	2	75
Namibia	16.50	14.48	5.24	16	8	8	–3	–3	55
Niger	0.73	0.33	0.15	12	47	47	–2	–2	43
Nigeria	2.37	1.92	0.41	25	30	27	–1	2	70
Rwanda	1.70	1.64	0.13	43	37	31	9	15	83
S. Tome and Principe	5.37	5.44	1.97	16	20	18	–10	–8	24

Internet (2002)					Population (2002)		Income
Users per 10.000 inhab.	CARG 98–02 (in%)	Hosts per 10.000 inhab.	PC per 100 inhab.	CARG 98–02 (in %)	Total, million	Rural pop. (% of total pop)	GN per capita, Atlas method (current US$)
29.42	94	0.01	0.19	23.72	13.90	65	660
73.52	96	0.84	0.22	17.12	6.60	56	380
290.70	46	9.40	4.07	12.37	1.71	50	2980
20.90	46	0.34	0.16	14.54	11.83	83	220
12.02	67	0.00	0.07	. .	7.07	90	100
37.90	128	0.28	0.57	23.47	15.52	50	560
364.46	66	1.09	7.97	60.60	0.46	36	1290
12.64	45	0.02	0.20	19.11	3.83	58	260
19.06	152	0.01	0.17	9.94	8.14	75	220
41.99	93	0.16	0.55	24.70	0.59	66	390
15.15	155	0.11	0.39	5.08	3.19	33	700
9.50	291	0.03	0.00	. .	53.80	n/a	90
54.58	67	2.67	0.93	19.71	16.77	56	610
35.64	34	0.06	0.69	32.47	0.48	50	700
22.61	128	2.16	0.25	. .	4.31	80	160
7.42	65	0.01	0.15	26.13	67.33	84	100
192.46	83	0.61	1.92	22.42	1.29	17	3120
182.22	73	4.14	1.38	43.62	1.38	68	280
78.43	124	0.14	0.38	16.03	20.07	63	270
45.66	186	0.33	0.55	12.58	7.74	72	410
39.90	98	0.16	0.00	. .	1.25	67	150
125.27	120	0.93	0.64	15.51	31.34	65	360
96.91	216	0.21	0.00	. .	2.09	71	470
3.09	69	0.03	0.00	. .	3.30	54	150
34.57	53	0.32	0.44	25.78	16.44	69	240
25.87	89	0.02	0.13	13.60	10.74	85	160
23.52	84	0.15	0.14	11.31	11.35	68	240
37.28	73	0.29	1.08	14.91	2.83	40	410
991.33	40	28.60	11.65	7.81	1.21	58	3850
. .	. .	0.00	0.00	. .	0.15	. .	. .
16.45	66	1.06	0.45	15.85	18.44	66	210
266.67	73	19.78	7.09	31.57	1.82	68	1780
12.77	156	0.10	0.06	18.95	11.54	78	170
34.98	88	0.09	0.71	3.85	132.78	54	290
30.60	124	1.51	0.00	. .	8.16	93	230
728.48	125	70.79	0.00	. .	0.15	52	290

(continued)

	Total Telephone Subscribers pr 100 inhab. (in%)								
	Sept 2003	2002	1995	CARG 95–02 (in %)	Sept 2003 Ranking	2002 Ranking	Change of Ranking (95–03)	Change of Ranking (1995–2002)	% of Mobile in Total (2002)
Senegal	9.10	7.72	0.98	34	13	14	1	0	71
Seychelles	92.24	82.25	17.48	25	1	1	0	0	67
Sierra Leone	2.06	1.82	0.37	25	33	29	–2	2	73
Somalia	1.80	1.33	0.17	35	35	37	8	6	26
South Africa	40.28	41.05	11.49	20	3	3	0	0	74
Sudan	3.25	2.65	0.28	38	26	24	11	13	22
Swaziland	10.11	9.50	2.32	22	11	11	–2	–2	64
Tanzania	3.25	2.41	0.33	33	25	25	8	8	81
Togo	4.68	4.54	0.52	36	23	21	0	2	77
Uganda	2.73	1.81	0.21	36	27	30	13	10	88
Zambia	2.28	2.12	0.88	13	32	26	–17	–11	61
Zimbabwe	5.80	5.51	1.42	21	19	17	–7	–5	55
WORLD AVERAGE	..	36.5	13.78	15	..	..	..	n/a	51.2
SSA AVERAGE	5.81	5.26	1.25	23	..	..	..	n/a	71.41
SSA AVERAGE Exc. SA	3.34	2.69	0.48	28	..	..	..	n/a	68.53

Note: (a) 2002 Data for Mainline in 3rd Q2003, (b) 2001 Data for Mainline in Liberia and Mayotte, (c) 2001 Data for Mobile in Liberia for 2002, (d) 3rd Q2003 Population based CAGR of period 1997–2002

Source: WB Analysis based on ITU World Telecommunications Indicators Database, Middle East and Africa Wireless Analyst for 2003 Data on Mobile and WDI.

Internet (2002)					Population (2002)		Income
Users per 10.000 inhab.	CARG 98–02 (in%)	Hosts per 10.000 inhab.	PC per 100 inhab.	CARG 98–02 (in %)	Total, million	Rural pop. (% of total pop)	GN per capita, Atlas method (current US$)
104.20	88	0.76	1.98	10.46	10.01	51	470
1452.10	55	32.91	16.08	7.49	0.08	35	6530
16.16	89	0.56	0.00	..	5.24	62	140
87.58	440	0.00	0.00	..	9.39	72	..
682.01	23	43.75	7.26	7.39	43.58	42	2600
25.82	146	0.00	0.61	33.35	32.37	62	350
193.80	107	12.88	2.42	..	1.09	73	1180
23.23	121	0.50	0.42	23.73	35.18	66	280
410.42	86	0.16	3.08	45.74	4.77	66	270
40.49	55	0.91	0.33	15.46	23.40	85	250
49.01	100	1.52	0.75	4.92	10.46	60	330
429.75	163	2.05	5.16	44.95	12.97	63	n/a
981.79	..	235.91	9.26	..	6,201.38	52	5080
93.04	40	3.48	1.05	12	687.99	67	450
50.66	91	0.58	0.68	17	655.63	..	..

Status of Telecommunications Sector Reform in Sub-Saharan Africa (end 2003)

	Separate Posts & Telecom	Incumbent Operator Privatized	Privatization Details	Separate Regulatory Authority	Competition in Mobile	Competition in international voice	Total penetration Sept 2003 in %	GNI per cap., Atlas method (current US$)
Angola	v	No	n/a	v	v	Partial	2.35	660
Benin	No	No	n/a	v	v	Monopoly	4.58	380
Botswana	v	No	n/a	v	v	Partial	34.65	2980
Burkina Faso	v	No	n/a	v	v	Monopoly	1.90	220
Burundi	v	No	n/a	v	v	v	1.11	100
Cameroon	v	No	n/a	v	v	Monopoly	6.73	560
Cape Verde	v	v	>51% private	v	Monopoly	Monopoly	27.38	1290
Central African	v	v	<50%	No	v	Monopoly	0.87	260
Chad	v	No	n/a	v	v	Monopoly	0.87	220
Comoros	No	No	n/a	v	Monopoly	Monopoly	5.08	390
Congo	v	No	n/a	v	v	v	9.31	700
Congo	No	No	n/a	v	v	v	1.65	90
Cote d'Ivoire	v	v	>50% private	v	v	Monopoly	8.68	610
Equatorial	v	v	<50%	No	Monopoly	Monopoly	8.47	700
Eritrea	v	No	n/a	No	No Operator	Monopoly	0.89	160
Ethiopia	v	No	n/a	v	Monopoly	Monopoly	0.61	100
Gabon	v	No	n/a	v	v	Monopoly	22.59	3120
Gambia	v	No	n/a	No	v	Monopoly	12.09	280
Ghana	v	v	<50%	v	v	Partial	4.74	270
Guinea	v	v	>50%	No	v	Partial	1.62	410
Guinea-	v	v	>50%	v	No Operator	Monopoly	0.88	150
Kenya	No	No	n/a	v	v	Monopoly	7.35	360
Lesotho	v	v	>50%	v	v	Monopoly	8.16	470
Liberia	v	No	n/a	No	v	v	1.34	150

Madagascar	v	>50%	v	v	v	1.79	240
Malawi	No	n/a	v	v	Partial	1.46	160
Mali	No	n/a	v	v	Partial	2.45	240
Mauritania	v	>50%	v	v	Partial	11.07	410
Mauritius	v	<50%	v	v	Monopoly	60.20	3850
Mayotte	..	..	..	..	..	20.84	n/a
Mozambique	No	n/a	v	v	Partial	2.37	210
Namibia	No	n/a	v	Monopoly	Monopoly	16.50	1780
Niger	v	>50%	v	v	Monopoly	0.73	170
Nigeria	No	n/a	v	v	Partial	2.37	290
Rwanda	No	n/a	v	Monopoly	Monopoly	1.70	230
S. Tome & Principe	v	>50%	No	Monopoly	Monopoly	5.37	290
Senegal	v	>50%	v	v	Monopoly	9.10	470
Seychelles	v	>50%	v	v	Partial	92.24	6530
Sierra	v	n/a	No	v	Partial	2.06	140
Somalia	v	100%	No	v		1.80	n/a
South Africa	v	>50%	v	v	Partial	40.28	2600
Sudan	v	>50%	v	Monopoly	Monopoly	3.25	350
Swaziland	No	n/a	No	Monopoly	Monopoly	10.11	1180
Tanzania	v	<50%	v	v	Partial	3.25	280
Togo	v	n/a	v	v	Partial	4.68	270
Uganda	v	>50%	v	v	Partial	2.73	250
Zambia	No	n/a	v	v	Monopoly	2.28	330
Zimbabwe	No	n/a	v	v	v	5.80	n/a
Number of v	43	20	37	36	7		

Source: WB Analysis based on ITU World Telecommunications Indicators Database, World Telecommunication Regulatory Database and East and Africa Wireless Analyst for 2003 Data on